FRONTIERS OF SCIENCE

BIOLOGICAL SCIENCES

FRONTIERS OF SCIENCE

BIOLOGICAL SCIENCES

Notable Research and Discoveries

KYLE KIRKLAND, PH.D.

BIOLOGICAL SCIENCES: Notable Research and Discoveries

Facts On File, Inc.
An imprint of Infobase Publishing
132 West 31st Street
New York NY 10001

Library of Congress Cataloging-in-Publication Data

Kirkland, Kyle.
Biological sciences: notable research and discoveries / Kyle Kirkland.
p. cm.—(Frontiers of science)
Includes bibliographical references and index.
ISBN 978-0-8160-7439-6
1. Medical sciences—Research. 2. Biology—Research. 3. Discoveries in science. I. Title.
R850.K45 2010
610.72—dc22 2009015651

Facts On File books are available at special discounts when purchased in bulk quantities for businesses, associations, institutions, or sales promotions. Please call our Special Sales Department in New York at (212) 967-8800 or (800) 322-8755.

You can find Facts On File on the World Wide Web at http://www.factsonfile.com

Text design and composition by Kerry Casey
Illustrations by Sholto Ainslie
Photo research by Tobi Zausner, Ph.D.
Cover printed by Bang Printing, Inc., Brainerd, Minn.
Book printed and bound by Bang Printing, Inc., Brainerd, Minn.
Date printed: February 2010
Printed in the United States of America

10 9 8 7 6 5 4 3 2 1

This book is printed on acid-free paper.

CONTENTS

PREFACE

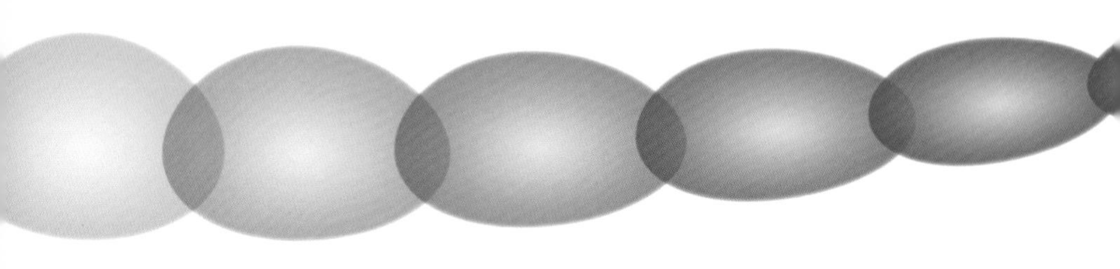

Discovering what lies behind a hill or beyond a neighborhood can be as simple as taking a short walk. But curiosity and the urge to make new discoveries usually require people to undertake journeys much more adventuresome than a short walk, and scientists often study realms far removed from everyday observation—sometimes even beyond the present means of travel or vision. Polish astronomer Nicolaus Copernicus's (1473–1543) heliocentric (Sun-centered) model of the solar system, published in 1543, ushered in the modern age of astronomy more than 400 years before the first rocket escaped Earth's gravity. Scientists today probe the tiny domain of atoms, pilot submersibles into marine trenches far beneath the waves, and analyze processes occurring deep within stars.

Many of the newest areas of scientific research involve objects or places that are not easily accessible, if at all. These objects may be trillions of miles away, such as the newly discovered planetary systems, or they may be as close as inside a person's head; the brain, a delicate organ encased and protected by the skull, has frustrated many of the best efforts of biologists until recently. The subject of interest may not be at a vast distance or concealed by a protective covering, but instead it may be removed in terms of time. For example, people need to learn about the evolution of Earth's weather and climate in order to understand the changes taking place today, yet no one can revisit the past.

Frontiers of Science is an eight-volume set that explores topics at the forefront of research in the following sciences:

- biological sciences
- chemistry
- computer science

- Earth science
- marine science
- physics
- space and astronomy
- weather and climate

The set focuses on the methods and imagination of people who are pushing the boundaries of science by investigating subjects that are not readily observable or are otherwise cloaked in mystery. Each volume includes six topics, one per chapter, and each chapter has the same format and structure. The chapter provides a chronology of the topic and establishes its scientific and social relevance, discusses the critical questions and the research techniques designed to answer these questions, describes what scientists have learned and may learn in the future, highlights the technological applications of this knowledge, and makes recommendations for further reading. The topics cover a broad spectrum of the science, from issues that are making headlines to ones that are not as yet well known. Each chapter can be read independently; some overlap among chapters of the same volume is unavoidable, so a small amount of repetition is necessary for each chapter to stand alone. But the repetition is minimal, and cross-references are used as appropriate.

Scientific inquiry demands a number of skills. The National Committee on Science Education Standards and Assessment and the National Research Council, in addition to other organizations such as the National Science Teachers Association, have stressed the training and development of these skills. Science students must learn how to raise important questions, design the tools or experiments necessary to answer these questions, apply models in explaining the results and revise the model as needed, be alert to alternative explanations, and construct and analyze arguments for and against competing models.

Progress in science often involves deciding which competing theory, model, or viewpoint provides the best explanation. For example, a major issue in biology for many decades was determining if the brain functions as a whole (the holistic model) or if parts of the brain carry out specialized functions (functional localization). Recent developments in brain imaging resolved part of this issue in favor of functional localization by showing that specific regions of the brain are more active during certain tasks. At the same time, however, these experiments have raised other questions that future research must answer.

The logic and precision of science are elegant, but applying scientific skills can be daunting at first. The goals of the Frontiers of Science set are to explain how scientists tackle difficult research issues and to describe recent advances made in these fields. Understanding the science behind the advances is critical because sometimes new knowledge and theories seem unbelievable until the underlying methods become clear. Consider the following examples. Some scientists have claimed that the last few years are the warmest in the past 500 or even 1,000 years, but reliable temperature records date only from about 1850. Geologists talk of volcano hot spots and plumes of abnormally hot rock rising through deep channels, although no one has drilled more than a few miles below the surface. Teams of neuroscientists—scientists who study the brain—display images of the activity of the brain as a person dreams, yet the subject's skull has not been breached. Scientists often debate the validity of new experiments and theories, and a proper evaluation requires an understanding of the reasoning and technology that support or refute the arguments.

Curiosity about how scientists came to know what they do—and why they are convinced that their beliefs are true—has always motivated me to study not just the facts and theories but also the reasons why these are true (or at least believed). I could never accept unsupported statements or confine my attention to one scientific discipline. When I was young, I learned many things from my father, a physicist who specialized in engineering mechanics, and my mother, a mathematician and computer systems analyst. And from an archaeologist who lived down the street, I learned one of the reasons why people believe Earth has evolved and changed—he took me to a field where we found marine fossils such as shark's teeth, which backed his claim that this area had once been under water! After studying electronics while I was in the air force, I attended college, switching my major a number of times until becoming captivated with a subject that was itself a melding of two disciplines—biological psychology. I went on to earn a doctorate in neuroscience, studying under physicists, computer scientists, chemists, anatomists, geneticists, physiologists, and mathematicians. My broad interests and background have served me well as a science writer, giving me the confidence, or perhaps I should say chutzpah, to write a set of books on such a vast array of topics.

Seekers of knowledge satisfy their curiosity about how the world and its organisms work, but the applications of science are not limited to intellectual achievement. The topics in Frontiers of Science affect society on a multitude of levels. Civilization has always faced an uphill battle to procure scarce resources, solve technical problems, and maintain order. In modern times, one of the most important resources is energy, and the physics of fusion potentially offers a nearly boundless supply. Technology makes life easier and solves many of today's problems, and nanotechnology may extend the range of devices into extremely small sizes. Protecting one's personal information in transactions conducted via the Internet is a crucial application of computer science.

But the scope of science today is so vast that no set of eight volumes can hope to cover all of the frontiers. The chapters in Frontiers of Science span a broad range of each science but could not possibly be exhaustive. Selectivity was painful (and editorially enforced) but necessary, and in my opinion, the choices are diverse and reflect current trends. The same is true for the subjects within each chapter—a lot of fascinating research did not get mentioned, not because it is unimportant, but because there was no room to do it justice.

Extending the limits of knowledge relies on basic science skills as well as ingenuity in asking and answering the right questions. The 48 topics discussed in these books are not straightforward laboratory exercises but complex, gritty research problems at the frontiers of science. Exploring uncharted territory presents exceptional challenges but also offers equally impressive rewards, whether the motivation is to solve a practical problem or to gain a better understanding of human nature. If this set encourages some of its readers to plunge into a scientific frontier and conquer a few of its unknowns, the books will be worth all the effort required to produce them.

ACKNOWLEDGMENTS

Thanks go to Frank K. Darmstadt, executive editor at Facts On File, and the FOF staff for all their hard work, which I admit I sometimes made a little bit harder. Thanks also to Tobi Zausner for researching and locating so many great photographs. I also appreciate the time and effort of a large number of researchers who were kind enough to pass along a research paper or help me track down some information.

INTRODUCTION

In 1676, Antoni van Leeuwenhoek (1632–1723) looked through his microscope at a drop of water and expanded the frontiers of biology in a dramatic way. Leeuwenhoek, a Dutch merchant whose name is difficult for English speakers to pronounce (most English-language speakers say "layvenhook" or "laywenhook"), learned how to grind optical lenses to magnify tiny objects. He built simple microscopes—instruments with a single lens—and examined the textiles he was selling. Then he turned his attention to other objects. He observed bee stingers and algae, among other objects, and began writing about his discoveries to the Royal Society of London in 1673. Three years later he saw tiny organisms in water and published his observations to skeptical scientists.

Before Leeuwenhoek's discovery, people knew nothing of bacteria and other microorganisms. Diseases such as cholera were well known, but no one realized that cholera was caused by bacteria in the water. It took a while for people to connect bacteria with diseases—the "germ" theory of disease did not become widely accepted until French scientist Louis Pasteur (1822–95) demonstrated in the 19th century the pervasiveness of microorganisms—but Leeuwenhoek, British researcher Robert Hooke (1635–1703), and others paved the way.

Expansion of knowledge by means of technology, such as with a microscope, is a common theme in biology, as it is in other sciences. *Biological Sciences: Notable Research and Discoveries,* one volume of the Frontiers of Science set, is about scientists who explore the frontiers of the biological sciences—and often find things they do not expect. Biology is the study of living organisms or processes involved in life; the term *biology* derives from a Greek word, *bios,* meaning life or mode of life, and *logos,* meaning word or knowledge. The biological sciences include a range of related disciplines—physiology, genetics, ecology, botany, molecular biology, and the study of specific biological systems such as the nervous system. The book discusses six topics that encompass a wide range of the biological sciences.

In Leeuwenhoek's day, knowledge of life and its mechanisms and processes was severely limited. Scientists of the 17th century viewed biology with a great deal of reserve due to its complexity—living organisms were clearly more complex than most inanimate matter. The subject of life also had a special status—humans are included in the subject matter—and many early scientists were uncomfortable with the prospect of possibly dehumanizing people by classifying them as objects to study. People of the 17th century tended to view life as the domain of special forces, such as vital spirits that somehow flowed through organisms to animate their actions. According to this old view, life was fundamentally static—although individuals changed and aged, the many types of life, such as plants and animals, stayed the same. These beliefs persisted well into the 18th century and beyond.

Yet technology, as well as the curiosity of researchers, spurred progress, and the pace is rapidly accelerating. In 1859, British biologist Charles Darwin (1809–82) outlined his theory of evolution, which proposed that variations enhancing the ability of organisms to survive and reproduce are passed from parent to offspring, causing species to adapt and evolve. It took 100 years for scientists to discover the molecular identity of these units of inheritance—*deoxyribonucleic acid* (DNA)—but only about 50 years passed after this discovery before scientists had mapped all of human DNA.

The benefits of this progress are immense. Scourges such as smallpox have been eradicated, treatments for diseases such as cancer and heart disease are improving, and scientists are accumulating important knowledge to help them understand and preserve Earth's essential ecosystems.

But there are still many frontiers in the biological sciences awaiting exploration. Each chapter of this book explores one of these frontiers. Reports published in journals, presented at conferences, and reported in news releases describe research problems of interest in the biological sciences, and how scientists are tackling them. *Biological Sciences: Notable Research and Discoveries* discusses a selection of these reports—unfortunately there is room for only a fraction of them—that offer the student and other readers insight into the methods and applications of biology.

The biological sciences can be complicated subjects. Students need to keep up with the latest developments in these rapidly advancing fields, but they have difficulty finding a source that explains the basic concepts while discussing the background and context essential for the "big picture." The book describes the evolution of each of the six main topics it covers, and explains the problems that researchers are currently investigating as well as the methods they are developing to solve them.

Chapter 1 describes how scientists who study the brain are discovering the functional roles of each part of this astonishingly complex system. Images of brain activity, which can now be produced from human subjects as they think and perceive, help researchers to correlate the activity of specific regions to the thought processes they create. As brain science advances, even the mysteries of human consciousness are being explored.

The influence of *genes* and genetic information is also critical for behavior, as well as for many types of diseases to which people are susceptible in varying degrees. To accelerate research in this field, scientists decided to read the human *genome*—the entire genetic material—through a huge effort called the Human Genome Project. Chapter 2 discusses how researchers are using this enormous amount of data to locate genes that cause disease and influence behavior—and also to identify people who may experience negative reactions to certain drugs.

Genes are the templates for *proteins,* and proteins are the workhorses of the body. Certain proteins catalyze chemical reactions, speeding them up so that they are fast enough to support the needs of the organism; other proteins transport cargoes, provide structural support, or become weapons against invaders. Chapter 3 explores how researchers are studying the shape of these molecules, and how this shape affects their many functions.

Other biological scientists have focused on change, variability, and the consequences of evolution. As a result of variability, Earth contains a diversity of organisms, as discussed in chapter 4. This diversity is critical in shaping life and the environment in ways that scientists have yet to fully understand. Researchers are using special molecules, carefully controlled environments, and sophisticated computer programs to study the relationship between diversity and the environment.

Biology is a wide-ranging discipline that can be difficult to define precisely because life is so variable—and can also sometimes be difficult to define. A *virus*, the subject of chapter 5, is a case in point. These tiny objects possess some of the characteristics of life, such as the ability to replicate themselves, but not others—they have no means of turning food into energy, for example. Many biologists do not consider viruses to be living organisms, but they are made of biological substances and they infect various forms of life, often causing serious diseases, so biologists study them.

Sometimes an unusual observation will spark a whole new branch of biology. When people noticed that salamanders can regrow a lost limb, they began to wonder how these remarkable creatures could do such a thing—and whether this process could be applied elsewhere to replace lost or damaged tissue in humans. But the mechanisms underlying these observations were mysterious, until scientists at the frontiers of biology began probing the hidden processes. The salamander research led to the study of *regeneration*, covered in chapter 6.

The discoveries of Leeuwenhoek, Darwin, Pasteur, and others have profoundly altered the way people think about life. Living organisms remain complex, but as biologists peer further into the molecular level, at proteins and DNA, or step back and take a global view of subjects such as biodiversity, life becomes more understandable.

Scientific knowledge also has tremendous benefits. Acceptance of the germ theory of disease, for instance, resulted in improved sanitation, sterilization of surgical instruments, and similar measures that have saved millions of lives over the years. Topics at the frontiers of biology, including the research described in each of the following chapters, have the potential for even greater benefits, as well as providing the satisfaction that comes with a better understanding of life and Earth's most complex organisms.

1

Brain Imaging: Searching for Sites of Perception and Consciousness

In 1924, German psychiatrist Hans Berger (1873–1941) found what he believed was a "brain mirror." Working at the University of Jena in Germany, Berger was studying a patient who had recently undergone a brain operation. Berger's initial effort focused on stimulating the brain by sending electrical current through the skull via special conductors called *electrodes,* which were attached to the patient's scalp. One day he unhooked the stimulator and connected the electrodes to a galvanometer. This instrument does not produce current but instead measures and records it. Physicians in that era often used galvanometers to record the electrical activity of the heart (this recording is called an electrocardiogram), but when Berger connected the scalp electrodes he saw squiggly lines representing brain activity.

Berger believed this recording, the *electroencephalogram* (EEG), could reflect or mirror the activity of the human brain. In 1929, after refining his equipment and conducting many more experiments, Berger began to publish his results. But other scientists were skeptical. The passage through the skull and scalp distorts the signal, and unrelated activity, such as that which comes from the muscles, makes unwanted contributions.

As a pioneer, Berger blazed a trail for others to follow, although his death in 1941 came before his work was duly appreciated. Instruments and recording techniques improved, and the EEG subsequently became an important tool in medicine and science. The EEG proved especially important in the study of abnormal electrical activity in the brain called seizures. Seizure disorders, also known as epilepsy, result when waves of electrical activity in part or in all of the brain become unusually synchronized (so that most of the brain is active at the same time), which often causes the patient to lose consciousness and experience uncontrolled muscular contractions. But despite its usefulness, the EEG is limited; in addition to the problems cited above, it does not generally allow pinpointing the origin of the recorded activity, and scientists realized they needed better methods to visualize brain activity. This chapter describes how modern scientists study the brain with much improved "mirrors" that help them discover the function of each part of the brain, and how these parts work together to create thoughts and minds.

INTRODUCTION

One of the most important frontiers of biology today is *neuroscience,* the study of the brain. (The prefix *neuro* comes from a Greek word, *neuron,* meaning nerve.) Biology is a mature subject but neuroscience is a relatively new discipline, growing prominent only in the 1960s. The delay in establishing neuroscience is surprising, considering the importance of the brain. Housed in the brain's three pounds (1.4 kg) of tissue is the basis for consciousness and memories, as well as the ability to coordinate the muscles and perform athletics—the brain does everything that makes a person unique and special.

Early biologists did not ignore the brain, but they could make little progress, since this organ is extremely difficult to study. Its activity is hidden by the skull, which protects the delicate tissue. Even when exposed, the brain offers little clue of its inner workings to the unaided human eye. In ancient times, the noted Greek philosopher Aristotle (384–322 B.C.E.) did not even believe the brain was important for behavior. Perhaps Aristotle based his mistaken belief on a peculiar observation—a chicken can still run around for a short period of time after its head is removed, suggesting that muscle activation does not require the brain. But observers such as Galen (129–99 C.E.), a Greek physician

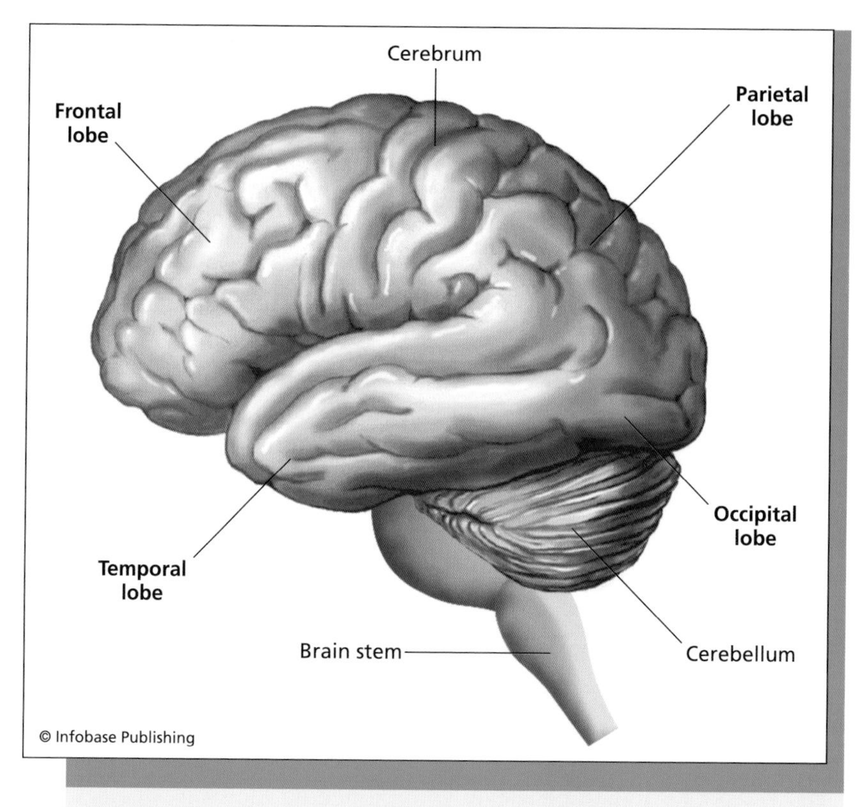

This drawing shows the four lobes of one of the two cerebral hemispheres of the human brain—the cerebellum and brainstem are also shown.

who treated gladiators in the Roman Empire, witnessed plenty of cases where injuries to a person's brain corresponded to deficits in movement, speech, perception, and thinking. For example, injuries to the back of the brain tend to be associated with vision problems. (As for the motion of headless chickens, this movement comes from activity in the spinal cord, which is normally under the control of the brain. Released from the brain's influence, the spinal cord may briefly issue a flurry of commands before the animal expires, resulting in a wild and eerie run.)

Anatomists went on to examine the structure of the brain and identify its components. The large anterior (front) portion of the brain is the cerebrum, as shown in the figure, and the posterior (rear) structure, tucked underneath the cerebrum, is the cerebellum ("little" brain). The cerebrum consists of two *cerebral hemispheres.* Each hemisphere has four main lobes—frontal, temporal, parietal, and occipital—that the 19th-century French anatomist Louis Pierre Gratiolet (1815–65) named for the adjacent bones of the skull. Covering the surface of the hemispheres is the *cerebral cortex.* (*Cortex* is a Latin word meaning bark, as in the outer covering of a tree.) The cortex of each lobe can be generally referred to by the name of the lobe; for example, cortex of the frontal lobe is called frontal cortex.

All life forms and their organs and tissues are based on the *cell.* Cells are small (usually with diameters of about 0.0004–0.004 inches [0.0001–0.01 cm] in size), filled with a water solution containing important molecules and nutrients, and surrounded by a lipid (fatty) membrane. Multicellular organisms such as humans are composed of many different kinds of cell, including a variety of blood cells, skin cells, liver cells, and many others. The brain consists of several cell types belonging to two main categories: glial cells, which support and nourish the brain, and neurons, which are the electrically active cells that generate the signals Hans Berger observed in his experiments. An adult human brain contains about one trillion neurons.

Long before Berger, scientists discovered the importance of electricity in the function of nervous systems. In 1791, Luigi Galvani (1737–98), an Italian physician who pioneered the study of electricity in biology, reported that electrical current in the nerves of frog legs made the muscles twitch. (Researchers named the galvanometer in honor of Galvani.) Soon thereafter scientists began probing the brain with electricity. Two German researchers, Eduard Hitzig (1838–1907) and Gustav Fritsch (1838–1927), showed in 1870 that certain areas of a dog's brain correspond to certain parts of the body. When the scientists electrically stimulated one small part of the cortex, a specific part of the dog's body moved. There was an area of the brain devoted to the rear legs, another for the fore legs, and so on, for each body part.

These electrical currents produce their effects by stimulating neurons. Embedded in neurons are proteins called *ion channels* that generate a brief impulse of electricity known as an action potential. The action potential proceeds down a long, thin section of the neuron called an axon, as shown in the figure. At the tip of an axon, the impulse causes the release of small membranous packets, called vesicles, filled with certain molecules. These neurotransmitter molecules drift across a small gap between the neurons known as a *synapse,* and

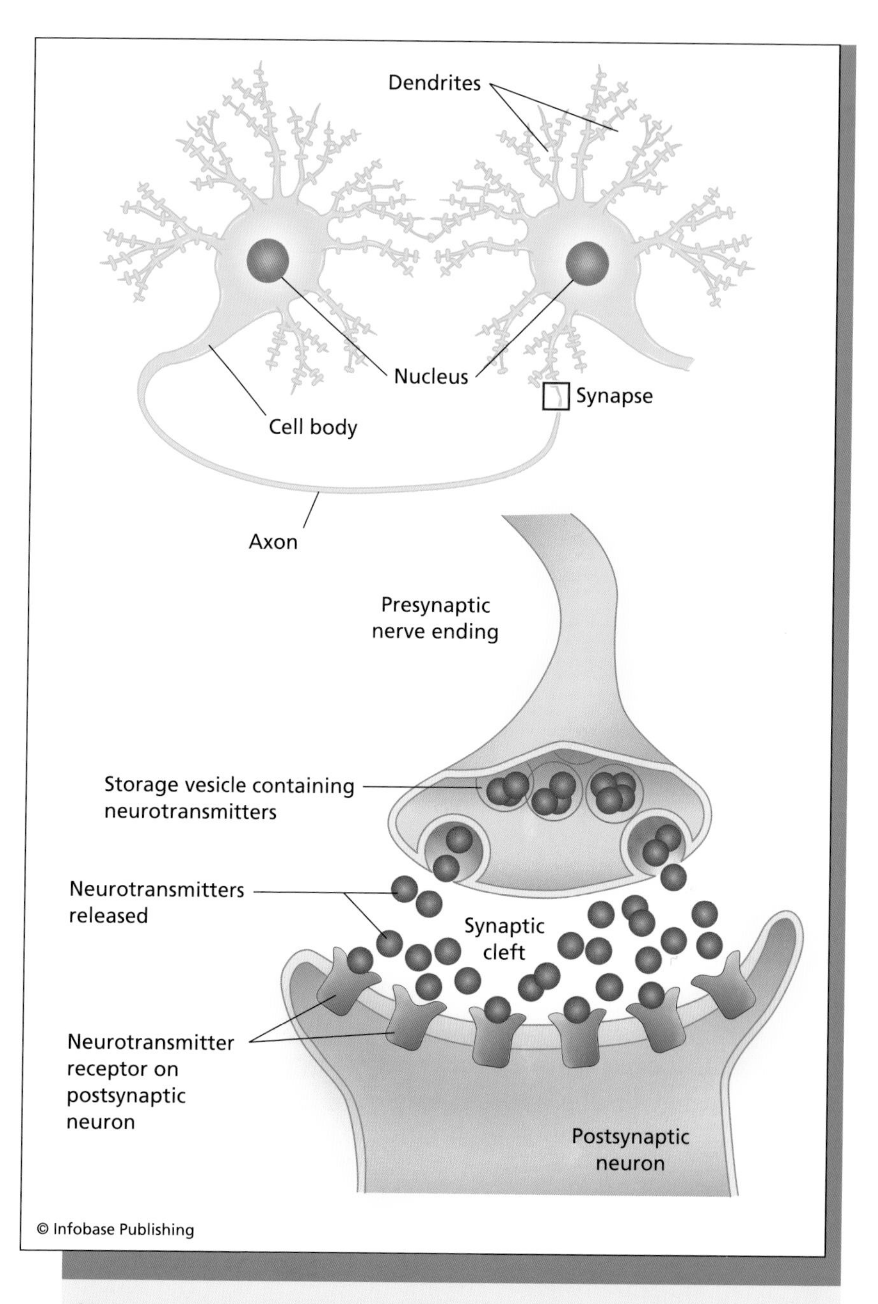

Neurons encode information in action potentials, which travel down the axon and initiate the release of neurotransmitters that bind to receptors in the recipient neuron. Some receptors are excitatory, increasing the chance that the recipient neuron will fire its own action potential, but some receptors are inhibitory, decreasing the chance.

usually act upon proteins known as receptors embedded in the membrane of other neurons. As a result, the recipient neuron may undergo an action potential, or it may be prevented or discouraged from doing so. In this manner, neurons send messages to one another, conveyed by the influence of neurotransmitters. Neurons connected together with synapses form *neural networks* that process information in the brain. Some neurons send messages to muscles instead of other neurons; axons of certain neurons travel to specific muscles and control their contractions (bundles of these axons make up a nerve).

Once scientists had identified the basic organization and operating principles, the next task was to understand how the brain uses these components to perform its functions. One of the main questions was whether the functions are localized. For example, does vision—such as seeing a yellow car traveling down the street—require the whole brain, or is this function served by a specific area or network?

PEERING INSIDE THE SKULL

To answer this question, scientists traced neural networks, identifying which regions of the brain are connected together via synapses. For instance, photoreceptor cells in the retina, at the back of the eye, make synapses with neurons called ganglion cells, which in turn send axons that project to (make synaptic connections with) neurons located in a region deep in the brain called the thalamus. Neurons in the thalamus project to neurons in a specific region of the cerebral cortex called V1, which is located in the occipital lobe. V1 projects to other areas in the cortex (as well as sending a projection back to the thalamus). Photoreceptor cells convert light entering the eye into a varying electrical current, carried by small particles called ions, and the ganglion cells, along with other cells, turn this signal into a train of action potentials that carry the information. Vision occurs when the neural networks in the cerebral cortex correctly interpret these impulse messages.

One of the most puzzling questions of neuroscience is how this interpretation occurs. There is also the question of how the activity of a bunch of neurons, which are individually nothing but a simple cell, is able to create something as amazing as the conscious sensation of vision—a picture in the "mind's eye." This extremely difficult question will be addressed later in the chapter. The first, slightly easier puzzle could be tackled if researchers had the ability to watch information flow through neural networks as a person views an object.

Hans Berger's EEG was one of the first means to do this. But this method suffers from a number of problems and limitations. The EEG signals measured from the surface of the scalp do not come from a single neuron, but instead come from many neurons whose activity combines to form the recorded waveforms. This is because an electrode pasted to the scalp covers a broad area, with many neurons contributing some of the current. Due to this effect, researchers have difficulty identifying the origin and nature of the signals. The only time a scalp EEG signal becomes easily interpretable is when many neurons are active at the same time, such as the synchronization of seizures, and during oscillations, described in a later section. In a normal brain the various neural networks carry on their own "conversations" and are out of synchronization with other networks. Physicians often use the EEG to identify and study the abnormal synchronization of seizures, but researchers studying normal activity are frustrated because too many different messages are smeared together. Sometimes researchers record an EEG from inside the brain or on the surface, which results in an improved signal but requires surgery to open the skull. And if the electrode is large, the signals will still come from a huge number of neurons.

An alternative to the EEG is to study single neurons. Scientists can do this by opening the skull and using hair-thin electrodes positioned near or inside the neuron. Experiments with laboratory animals provide this opportunity, and beginning in the late 1950s two American researchers, David Hubel and Torsten Wiesel, recorded from single neurons in the thalamus and cerebral cortex of an anesthetized cat. Anesthesia acts on the brain to render an animal or person unconscious and, of course, affects the brain in the process, but the cat's visual system remained intact (although some of its functions were no doubt altered). The scientists displayed images on a screen in front of the cat's eyes and recorded the activity from single neurons as the cells processed the information. These experiments, which have subsequently been performed on many different animals and on all the sensory systems (hearing, touch, taste, and smell, in addition to vision), showed that neurons break down the sensory information into basic elements. In the case of vision, the elements include boundaries (for example, lines that form the outline of objects or separate one object from another) and color.

Recording from single neurons allows researchers to learn exactly what that neuron contributes to the processing of information. But these experiments do not reveal how the network as a whole functions. And because of the invasive nature of the experiments—the brain must be exposed—the subjects generally must be limited to laboratory animals.

Neuroscience experiments such as those described above are analogous to an effort to understand what is happening during a game by listening to the fans. Investigators who position a microphone next to the stadium can get a general idea of how the game is going from the roar of the crowd. This "experiment" is analogous to the EEG. Investigators who attach a microphone to one of the fans can record how one single individual is responding, but this information reflects only that person's viewpoint, an "experiment" that is analogous to single neuron recordings. What neuroscientists needed was a way to peer inside the skull and watch the whole game.

A perfect technique that provides a comprehensive view of the brain in action does not yet exist. But neuroscientists have developed a number of techniques today that are improvements on the EEG. Of the three techniques described in this chapter, two are based on *metabolism*—chemical reactions occurring in cells—and one makes use of magnetic fields.

Positron emission tomography (PET) detects high-energy photons of light created when positrons and electrons meet. A positron is the anti-matter particle to the electron. When the two meet they annihilate one another, producing a pair of photons called gamma rays that travel in opposite directions. Positron emission occurs when certain radioactive substances decay and emit, or give off, particles such as positrons. A positron cannot survive long in the presence of matter since it will eventually encounter an electron and become transformed, along with the electron, into a pair of oppositely moving photons. PET machines detect these photon pairs and create a three-dimensional image of their points of origin, a process called tomography. The point of origin is the place where the positron and electron met.

Only certain radioactive nuclei such as fluorine-18 and oxygen-15 emit positrons during decay. These nuclei can be produced by high-energy collisions in machines called cyclotrons, many of which are owned and operated by hospitals and research institutions. Researchers incorporate these radioactive atoms into molecules such as glucose, a sugar that the body breaks down (metabolizes) to yield energy. When injected into a test subject, the radioactive molecules accumulate in areas of the body that use the most energy; one of these regions is the brain, which possesses only 2 percent of the body's weight but accounts for 20 percent of the body's energy usage.

Radioactivity is dangerous because the emissions can generate heat and damage vital molecules including deoxyribonucleic acid (DNA), but only small, safe amounts are injected into the test subject's body. PET machines began appearing in the 1970s for a wide variety of medical and scientific imaging, and in the late 1980s Marcus Raichle, a professor at Washington University, and his colleagues began to use this technique to study the brain.

About the same time as PET appeared, a tool called *magnetic resonance imaging* (MRI) began supplementing the use of *X-ray* devices to image a patient's body. X-rays are high-frequency electromagnetic radiation that normally passes through the body, but the relatively heavy atoms of calcium in the bones absorb these frequencies. Physicians check bones for fractures by examining the X-ray "shadow" on a special film that is sensitive to X-rays. The softer structures of the body, such as internal organs, contain mostly lighter atoms such as hydrogen, carbon, and oxygen, and do not show up well in X-ray images.

MRI creates images by placing the body in a strong magnetic field and subjecting it to radio waves, which are also electromagnetic radiation but of a much lower frequency than X-rays. The radio waves interact with hydrogen atoms in the body, causing them to spin (resonate) in a certain direction and frequency. When the radio waves are turned off, the atoms return to their normal state, emitting energy that is detected by the MRI machine. Mapping these energies creates a detailed view of any tissue in the body that contains hydrogen. Since there are two atoms of hydrogen in water (H_2O) and the body is about 65 percent water by weight, most organs and structures can be imaged, including the brain. Physicians use MRI to inspect the body for tumors and other diseased tissue, and neuroscientists use MRI to study the anatomy of the brain by safely imaging a living subject.

But to study the function of the brain instead of just its relatively constant anatomy, MRI needs to be modified to yield a set of images showing the brain's activity. This is what *functional MRI* (fMRI) does. The technique employs MRI technology to track the brain's blood flow. Blood contains a weakly magnetic protein molecule called hemoglobin;

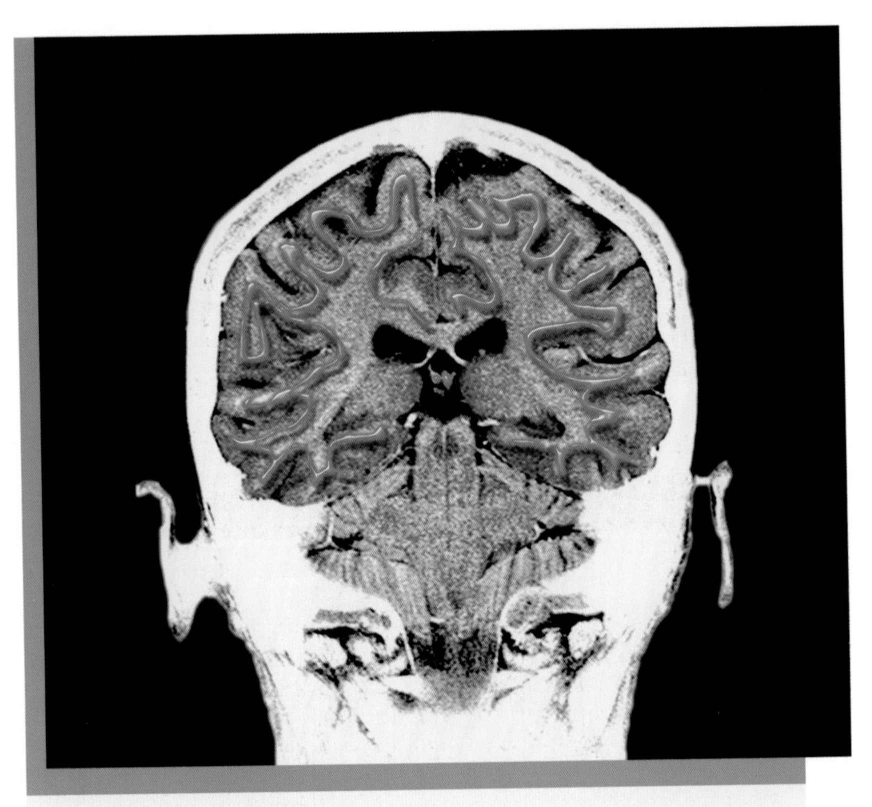

MRI image of a cross section of the human brain *(Living Art Enterprises, LLC/Photo Researchers, Inc.)*

oxygen molecules ride this protein as the blood circulates through the body, carrying needed oxygen to the cells. Hemoglobin's magnetic properties differ when oxygen is attached, and fMRI uses this difference to detect the flow of oxygenated blood through the tissue, which depends on how much energy the region is consuming. As neurons become more active they use more oxygen, so the oxygenation level dips. But shortly afterward the blood flow increases in response, boosting the oxygenation level. Researchers are not certain what mechanism causes this increase in blood flow with neural activity, but in any case, blood oxygenation provides a measurable though indirect signal of brain activity. Scientists began using fMRI for brain imaging in the 1990s.

An advantage of fMRI over PET is that it requires no injection of radioactive material. Although PET scans use only a small, safe dosage, subjects would be exposed to an unhealthy accumulated dose if scanned too often over a short period of time. There is also the trouble of obtaining the radioactive material.

But both fMRI and PET machines are expensive. An fMRI machine can cost $4 million or more, depending on the model, and a PET scanner runs about $2 million.

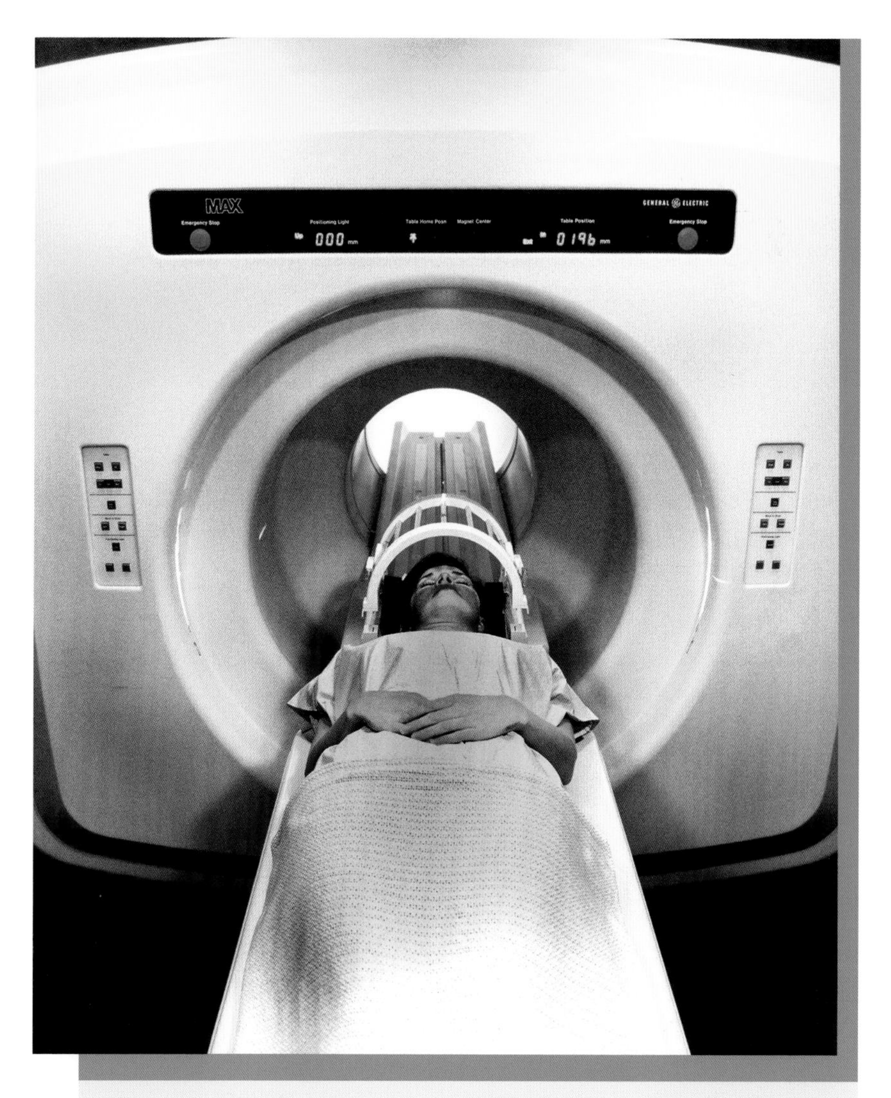

Patient entering an MRI scanner *(Charles Thatcher/Getty Images)*

Brain Imaging and Metabolism

In the late 19th century, British scientist Sir Charles Scott Sherrington (1857–1952) and his colleagues proposed that active brain cells cause changes in blood flow and blood oxygenation to the brain. This makes sense because the blood carries nutrients, such as glucose and oxygen, needed by the cells in order for them to generate energy. The production of action potentials is expensive in terms of energy—action potentials require the flow of ions across a neuron's membrane, and the ions must be pumped back or the neuron will lose its ability to produce more action potentials. Some neurons generate action potentials at rates of up to several hundred per second.

The idea motivating the use of PET and fMRI in neuroscience is that active brain regions need more energy. If one part of the brain participates in a specific function, then this part must be active while a person is performing the given function. For instance, when a person inspects an image, the visual system is active, which means that the parts of the thalamus and cerebral cortex involved in vision will need an extra supply of nutrients. Images created by PET and

PET and fMRI machines respond to metabolism—chemical reactions involved in energy flow—rather than the electrical activity of the brain, which is what an EEG records. But these newer techniques provide a three-dimensional map of metabolic activity of the brain, allowing researchers to pinpoint activity even deep in the brain. The usefulness of these techniques relies on the correspondence of the electrical activity of the brain to its energy requirements, as discussed in the above sidebar.

While some imaging techniques measure metabolic activity, *magnetoencephalography* (MEG) records the magnetic fields created by the fMRI indirectly measure the electrical activity of the brain by revealing the amount of fuel needed for the process. PET detects the accumulation of molecules that provide the energy, while fMRI detects changes in blood oxygenation levels.

A strict correspondence between electrical activity and energy consumption must not be assumed, however. Nikos Logothetis, a researcher at the Max Planck Institute for Biological Cybernetics in Tübingen, Germany, managed to make a direct measurement of electrical activity at the same time as obtaining an fMRI image. This electrical measurement, reported in the article "Neurophysiological Investigation of the Basis of the fMRI Signal" in a 2001 issue of *Nature,* is complicated because the strong magnetic fields of the fMRI tend to disrupt electrical equipment and probes. In a careful series of experiments using a specially designed magnet, Logothetis showed a strong relationship between electrical activity and the fMRI image (of an experimental animal), although the image reflected more of the inputs—the projections to the region—rather than the neural activity of the region itself. Marcus Raichle, writing in the same issue of *Nature,* described the result as "an experimental tour de force that represents the first comprehensive look at the relationship between the fMRI signal and the underlying neural activity."

tiny currents circulating in active neurons. These fields have exceptionally small magnitudes and require sensitive detectors such as superconducting quantum interference devices, which employ the principles of advanced physics. Shielding is necessary so that interference from other magnetic fields does not overwhelm the desired signals; for example, Earth's magnetic field, which affects compass needles, is about one billion times stronger than the brain's field. Although the measurements are difficult, the procedure offers a high-quality image of neural activity, and exceeds fMRI and PET in time *resolution*—the ability to show the time course of changes in activity.

All of the newer imaging techniques yield more information than EEGs and single neuron recordings. These techniques are far from perfect, for they are indirect measurements and are difficult to make, sometimes leading to errors in the elaborate analyses required to interpret the resulting images. Yet the techniques offer windows or mirrors by which neuroscientists can view brain activity that would otherwise be concealed, and were one of the reasons why, in 1990, President George H. W. Bush proclaimed the 1990s to be the Decade of the Brain. In Presidential Proclamation 6158, issued on July 17, 1990, to promote neuroscience research, Bush wrote, "Powerful microscopes, major strides in the study of genetics, and advances in brain imaging devices are giving physicians and scientists ever greater insight into the brain."

LOCALIZATION OF FUNCTION

By the time Hans Berger began his pioneering EEG studies in the 1920s, scientists had some crude notions about which parts of the brain did what. Researchers had identified "sensory areas" devoted to processing sensory information such as light and sound, "motor areas" that coordinated muscular contractions and movement, and "association areas" that were apparently for higher level functions. This knowledge of brain function came from stimulation experiments as well as studies of the behavioral, sensory, or motor deficits displayed by patients with brain injuries.

But concepts of motor, sensory, and association areas lacked specificity. In addition, brain science in its early days suffered from being linked with a peculiar pseudoscience—any subject in which practitioners misuse or misunderstand scientific concepts. As discussed in the following sidebar, promoters of the pseudoscience known as phrenology believed too much in specificity and had no experimental evidence for their conclusions. (The term *phrenology* derives from the Greek words *phrenos,* meaning mind, and *logos,* meaning word or knowledge.) Phrenologists claimed that bumps on the skull revealed a person's personality and aptitudes; the bumps were presumably the result of an enlarged development of the underlying region of brain tissue, which supposedly augmented a person's ability to perform whatever function this brain tissue served. For example, phrenologists informed people who had a bump on a specific region at the side of the head that they possessed an

Phrenology—"Reading" the Bumps of the Skull

Besides misleading a lot of people, an unfortunate result of phrenology was the sullying of the reputation of a careful and reliable scientist. Austrian anatomist Franz Joseph Gall (1758–1828) studied cranial nerves and the anatomy of the cerebral cortex. In 1808, he began promoting a theory that small, localized regions of the brain served specific mental faculties—the forerunner of modern ideas of localization of function. By examining the skulls of relatives, friends, and other people whom he knew, Gall tried to find correlations between skull bumps and mental faculties.

Gall's methods were scientific and his claims were generally modest and reserved. But the memory and reputation of this scientist became forever intertwined with people who followed in his footsteps and were much less careful, even abandoning science altogether. Subsequent phrenologists were not interested in making scientific discoveries; they were intent on creating a carnival-like sideshow in which credulous people, lacking scientific training, would pay a fee to get a "scientific" analysis of their individual strengths and weaknesses. The number of functions blossomed into a huge assortment of traits that included spirituality, conscientiousness, and combativeness. The figure illustrates an example of a phrenology map in which traits were assigned to specific regions of the head.

Instead of initiating a thriving new science of the brain, phrenology smothered brain science by misleading, misguiding, and otherwise obscuring the subject. Few knowledgeable people of the era put any stock in phrenology, but it was so widespread that scientists who studied localization of brain function risked losing credibility among their peers by being associated with this pseudoscience. Although the initial basis of phrenology—

(continues)

(continued)

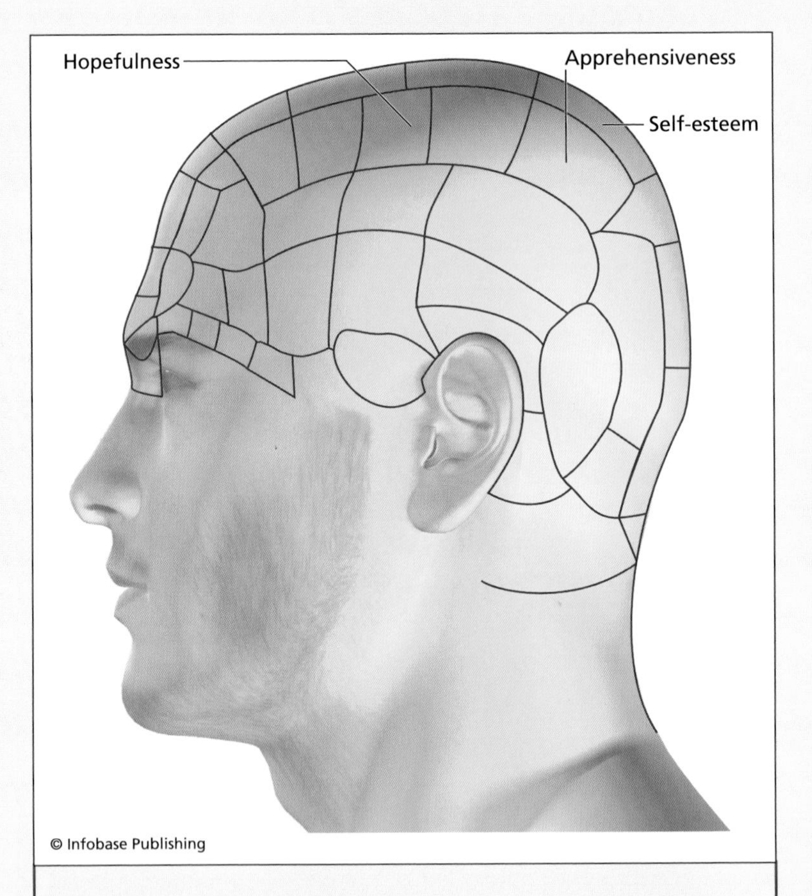

An example of a phrenology map, showing a few labeled areas. Practitioners believed that a prominent bump in a specific area of the skull meant the person possessed an abundance of the corresponding attribute.

Gall's theory that functions could be localized in the brain—was valid, the subject veered off in an unscientific direction that strangled advances in brain science for many years.

elevated sense of hopefulness; people without these bumps were held to be naturally gloomy in disposition. This extreme view of localization, unsupported by scientific experiment, was prominent in the middle of the 19th century and lingered for decades afterward.

After phrenology finally dissipated in the late 19th century, researchers took up in earnest the scientific study of functional localization. Debates focused on a central issue: Could functions such as language and memory be located in specific areas of the brain, as Gall theorized, or does the brain operate holistically—as a whole, with each part making a contribution to each function? Animal experiments suggested that in certain cases the brain operates holistically, but in other cases there are regions specifically devoted to certain functions such as vision and learning.

Physicians who studied patients with brain injuries uncovered strong evidence of functional localization. One of the pioneers of these studies was Paul Broca (1824–80), a French physician. Broca identified a specific region in the left hemisphere that correlated with a speech impediment. When damaged by some sort of lesion (injury), such as a stroke in which a certain amount of brain tissue dies from a lack of blood flow, the patient lost the ability to speak. Other researchers assigned functions to various regions of the brain in the same manner. By identifying behavioral, perceptual, or motor deficits of patients with brain lesions, researchers assumed that the particular region that had been damaged was normally responsible for the lost or impaired function.

There were many problems with these studies. Some of the deficits exhibited by patients were subtle and difficult to characterize, and physicians had no means of precisely locating which region of the brain had been damaged until after the patient expired, at which time an autopsy could be performed. The assumption that the lost function must normally be performed by the damaged area is also open to criticism, since the damage may have cut the flow of information instead of damaging the area in which the information is processed. Suppose, for example, a person cuts the wires that travel from a computer's memory to its central processing unit (CPU). The computer will be unable to perform computations, but this does not mean the wires perform computations—the wires merely carry the information necessary for the CPU to do its job.

Imaging tools such as PET, fMRI, and MEG eliminated these problems. Scientists could begin to conduct safe experiments on living subjects, obtaining the results quickly and easily. Unlike animal experiments, the subject is a human, so there is no issue of whether the findings apply to humans or not. Imaging tools also give researchers much more control over what function to investigate than they had with lesion studies.

Imaging experiments have often confirmed what earlier studies suggested. In the case of Broca's area, imaging studies have shown it is involved in speech, as he had suggested. Other areas that have been previously identified as important for speech also show up as highly active during the experiments, such as a region in the temporal lobe named after its discoverer, German physician Carl Wernicke (1848–1905). The images have been extremely useful in pinpointing and circumscribing these regions, which could only be roughly located in earlier studies.

For most people, one hemisphere is dominant for language—most of the functions of language are carried out in one or the other hemisphere—and in the majority of people this hemisphere is the left one. About 95 percent of right-handed people use the left hemisphere for language; for the other 5 percent, either the right hemisphere is dominant or, in some cases, neither hemisphere dominates. A lower percentage—60–70 percent—of left-handed people use the left hemisphere. (Researchers can study language dominance with the Wada test, named after Canadian physician Juhn Wada. The procedure anesthetizes only one hemisphere by injecting an anesthetic into that hemisphere's main artery. If the anesthetized hemisphere is dominant for language, the subject temporarily loses most of his or her ability to speak or understand language.) Neuroscientists do not yet know why language in humans is usually lateralized—performed in only one of the hemispheres—nor can they explain the differences between right- and left-handed people.

The data from brain imaging experiments have also unveiled many more areas that contribute to a person's use and understanding of speech. In the report "Human Brain Language Areas Identified by Functional Magnetic Resonance Imaging," published in 1997 in the *Journal of Neuroscience,* Jeffrey Binder, of the Medical College of Wisconsin, and his colleagues used fMRI to image the brains of 30 subjects while the subjects listened to spoken words. In addition to previously identified regions, Binder and colleagues discovered prominent activations in the frontal cortex. The cortex in this area receives inputs from many other areas and may play a role in decision-making and consciousness.

Imaging experiments allow not only the identification of areas generally involved in performing a task, but are also useful in analyzing the separate components that comprise the task. For example, the words of a language belong to different categories and serve different roles, such as nouns to specify objects and verbs to specify action. Researchers can use imaging to determine if any differences exist in where these categories are processed in the brain. Harvard University researchers Kevin A. Shapiro, Lauren R. Moo, and Alfonso Caramazza took fMRI images of the brains of people who produced either verbs or nouns in short phrases. As reported in "Cortical Signatures of Noun and Verb Production," published in *Proceedings of the National Academy of Sciences* in 2006, these researchers discovered that two areas of the brain were activated more strongly during verb production—a portion of the left frontal cortex and a region in the left parietal lobe. Noun production involved higher activation in a region of the left temporal lobe.

Experiments such as these help neuroscientists to construct a "functional neuroanatomy" or brain mapping that associates a specific function with a specific region or anatomical structure of the brain. Imaging studies have affirmed that functional localization exists in the human brain.

But interpretation of image experiments is not as easy as one would like. Complicated tasks consist of many different components, each of which invokes activity in a number of brain regions. Even a task as seemingly easy as reading engages an enormous number of areas—images of a person who is reading show activation in about 80 percent of the brain. Some of the activated networks are involved in vision (seeing the words on the page), some are involved in memory (remembering the word definitions), some are involved in interpreting the context and meaning of the materials, and some retrieve associations—the words may serve as a fragment that calls up an entire set of memories. (For instance, any time the author of this book reads the word *fragment,* he thinks of the 10 days he spent as a young serviceman on the Hawaiian island of Kahoolawe picking up bombshell fragments, sweating profusely, and hoping to avoid any encounters with a live shell.)

Widespread activation presents challenges to neuroscientists who design imaging experiments. Separating the components of a task or function often entails comparing images for a series of progressively more difficult subtasks. For example, in the 1997 experiment by Binder and colleagues described above, the scientists wished to distinguish between brain activation due to the processing of speech and activation that was involved only in the processing of sounds. To do this, the experimenters subtracted activation while the test subjects listened to nonlinguistic sounds from the activation obtained while the subjects heard words. Since the task of processing language included the task of processing sounds, subtracting the latter leaves the activation that was required over and above the basic chore of hearing and interpreting a sound—in this case, processing language.

PERCEIVING THE WORLD

As neuroscientists began to study the function of a variety of brain regions, they also began to realize that single brain regions do not work in isolation. The brain has much interconnectivity—neurons in many different regions communicate with each other, or are connected together via one or more other networks. The extreme localization of phrenology, in which small regions of the brain are wholly responsible for a complex trait such as intelligence, is not widely held today.

Although learning the function of certain areas and neural networks has been a great advance, this knowledge still does not fully answer the deeper question of how the brain works. To tackle this question, neuroscientists must explore how the different networks of the brain work together to produce perception—the mental images people obtain from their senses.

Each sensory system has its own networks, and the information for each system flows down different paths. Vision, hearing, olfaction (smell), taste, and touch have specific cells that detect the appropriate stimulus; for example, the eye contains photoreceptors to convert light falling on the retina into electrical signals. The ear contains special cells to convert sound into electrical signals, the nose has olfactory receptors to detect specific molecules, taste buds in the mouth detect chemicals in food, and special cells in the skin detect vibration or pressure. The signals of each of these sensory systems take a different route through the brain, and each has its own special processing centers in the cerebral cortex, devoted to analyzing the information of one specific sense.

Vision is the most widely studied sense, in part because of the important role it plays in many human activities. Reflecting its importance, more than half of the human brain contributes in some way to the processing of visual information. Through a combination of different types of experiment—lesion studies, single neuron recording in cats and monkeys, EEG, and imaging—scientists have traced the path of visual signals as they travel through the brain. The retina and subsequent regions on the pathway are structurally organized to maintain position information, so that the direction and location of an object can be determined. In the cortex, where the most advanced processing occurs, experimenters have found more than 30 distinct regions in monkeys that act on some part of these signals. The human visual system is similarly organized.

These areas of cortex involved in vision are arranged in tiers, or stages, with one set of regions connected with the next, and so on, as the information gets processed. Two main "streams" or pathways exist: one stream, which is located mostly in the lower (ventral) portion of the hemispheres, chiefly processes color and shape information; the other stream, which is located in the upper (dorsal) portion of the hemispheres, processes motion. The ventral stream's color analyzers include an area called V4. Imaging studies show that colorful images activate V4, and damage to this region causes difficulties in identifying colors. Regions that analyze the shape or form of objects also belong in the ventral stream. The dorsal stream contains regions such as V5 that are strongly activated by moving images.

Some researchers have identified regions of the cortex that appear to be used for highly specialized purposes. For example, Nancy Kanwisher, a professor at Massachusetts Institute of Technology, and her colleagues have proposed that a particular area of cortex called the fusiform face area specializes in identifying and recalling faces. (The fusiform cortex is situated on the underside of the temporal lobe, so it is part of the temporal cortex. Its name comes from its shape—the term *fusiform* refers to a rod or cylinder that is wide in the middle and small at the ends. The fusiform cortex is a component of the ventral stream of vision.) Social interaction among humans relies to a large extent on faces—people observe faces while spotting a relative or friend in a crowd, and facial features offer clues in determining a person's current state of emotion, such as anger, sadness, or elation. Brain imaging indicates

The human brain processes visual information such as color and motion along separate pathways, even though a brightly colored moving object invokes a single, unified perception. *(John Prescott/iStockphoto)*

that tasks involving facial recognition strongly activate the fusiform face area. Other studies show that patients with brain lesions limited to this small area are unable to recognize their friends and family, even though their vision is otherwise normal.

But notions of such highly specialized areas are difficult to prove beyond doubt. Although imaging experiments clearly demonstrate that most functions are localized to some extent—the functions do not require the whole brain—the widespread activity associated with behaviors such as reading suggest the possibility that many areas make contributions. Further experiments to analyze the contributions of each region are ongoing.

Localization of function requires the brain to distinguish various types of information and route the information appropriately through the various processing stages. In vision, for example, motion information travels along the dorsal stream and shape information gets processed in the ventral stream. The brain maintains this separation all the way to the end: There does not seem to be any single area that receives the output of all the other regions. In other words, there is no single cortical area that gets the "big picture." This is an extremely puzzling aspect of brain function. A person perceives an object as a whole, yet each part has been processed and analyzed in separate regions that do not send their "report" to a single region of the brain.

For example, suppose an observer sees a red car traveling down the street. To the observer's perception the car is a single object, yet each part of this image—the red color, the shape and identity of the car, and its motion—has been processed in separated regions of the brain. Somehow these regions work together to form a conscious awareness of a red car traveling down the street. How this happens is one of the central questions in the study of consciousness.

IMAGING THE MIND: NEURAL CORRELATES OF CONSCIOUSNESS

Imaging techniques discussed in this chapter have advanced the frontiers of neuroscience in identifying which networks and areas in the brain contribute to various functions such as language and perception. But now the frontier has reached exceptionally difficult problems, such as the nature of consciousness. How does a group of neural networks work together to give rise to a rich mental life and the "mind's eye"?

Several hypotheses have emerged. One hypothesis involves the coordination of brain activity by some sort of controller—a network or region of the brain that paces or supervises information processing across multiple areas. Although there is no single network in the brain to which all streams of information flow, the brain is highly interconnected and a number of networks receive synaptic input from a broad spectrum of other areas. The hypothetical network that would guide or spark consciousness may act as sort of a filter, or perhaps it may work as a spotlight to specify the center of attention. Perception is generally limited to one thing at a time—a person who is watching a red car traveling down the street usually cannot concentrate on anything else until the car loses the person's attention—and this hypothetical network may

Salk Institute for Biological Studies

Dr. Jonas Salk (1914–95) had a vision—an institute where scientists could work together and share their expertise in solving the most basic problems in biology. Salk was no stranger to ambitious goals; in 1955 his *vaccine* for polio, a viral infection responsible for paralyzing hundreds of thousands of people, proved effective and eliminated a major threat to the health of children across the world. In 1960, bolstered by support from the March of Dimes, Salk and his colleagues approached the San Diego City Council with a proposal for a research institute, if the council would provide the land. Recognizing the benefits of scientific research, the council agreed, a decision affirmed by the citizens in a referendum in June 1960.

The Salk Institute began with just a few members, but today the scientific staff numbers about 850, with nearly 60 faculty investigators. The major areas of study include molecular biology, genetics, neuroscience, and plant biology. Many young scientists come to the Salk Institute for training. Five of these scientists have gone on to win Nobel prizes.

Until recently, the faculty included the Nobel laureate Francis Crick, who passed away on July 28, 2004. Crick exemplified the well-rounded scientist, which Salk envisioned as the ideal member of the institute. Born in England and trained in physics and biology, Crick won the Nobel prize in physiology or medicine in 1962 with James Watson (1928–) and Maurice Wilkins (1916–2004) for discovering the structure

play an important role in selecting the object of interest and coordinating the information concerning it.

A number of possibilities for such a network or area exist. One of the leading candidates is the cortex at the anterior of the frontal lobe,

View of a portion of the Salk Institute campus *(Thomas A. Heinz/ Corbis)*

of DNA. Crick then went on to study problems in a variety of other sciences. One subject that intrigued him was the brain, and at the Salk Institute, which he joined in 1976, Crick taught himself neuroscience. With his active imagination, disciplined by scientific accuracy, he published papers on vision, learning and memory, and the neural basis of consciousness.

called the prefrontal cortex. This cortex is highly developed only in humans; other animals have little or no corresponding tissue in this region. The prefrontal cortex receives projections from a large number of other regions, sampling a wide variety of information pathways. Because of this, many neuroscientists believe the prefrontal cortex is involved in decision-making and solving problems. Imaging experiments show activation of this region during a variety of tasks requiring the subject to make a decision, based on an evaluation of a given set of instructions or stimuli.

Two theorists recently proposed another candidate, the claustrum. In a paper titled "What Is the Function of the Claustrum?" published in *Philosophical Transactions of the Royal Society B* in 2005, Francis C. Crick (1916–2004) of the Salk Institute for Biological Studies and Christof Koch at the California Institute of Technology discussed this as yet little known area, which lies hidden beneath the cerebral cortex at the side of the head. (*Claustrum* is Latin for enclosed space; this word is also the basis for the term *claustrophobia,* the fear of enclosed spaces.) No one is sure what the claustrum does, but it receives inputs from many areas of the cortex and sends a projection back to them. This interesting hypothesis formed one of the last papers of the late Francis Crick; the sidebar on page 24 highlights the Salk Institute, where the Nobel laureate made his many contributions to neuroscience.

Other regions of interest are the thalamus and the cortical areas to which it projects. Specific areas of the thalamus receive inputs from each type of sensory receptor, and convey this information to specific areas of the cortex (such as V1 in the visual system). The importance of these early processing areas in consciousness has been highlighted in phenomena such as blindsight.

Blindsight is a fascinating though tragic instance of a person who can respond to a visual stimulus despite being blind and unable to describe the stimulus. Patients exhibiting blindsight have damage to parts of their cortex that includes the primary visual cortex (V1). These patients are not consciously aware of objects located in the field of vision served by the damaged cortex. As described above, networks of the visual system maintain a map of the visual field; each hemisphere processes one half of the field—the left hemisphere processes the right field of vision and the right hemisphere processes the left half, since for as yet unknown reasons the information pathway crosses over. If, for example, the entire V1 region of the left hemisphere suffers a lesion, the patient will be blind in the right half of their field of vision. But when told about the presence of an object, a patient with blindsight can sometimes accurately reach for the object even though the patient cannot see it.

Blindsight is rare and difficult to study, so researchers have not yet determined the exact mechanisms that generate this unusual ability. What is apparent is that at least a few visual system pathways in some part or parts of the brain are spared damage in these patients. Although these pathways can guide the movement of the patient's hand, they cannot create the visual perception of the object.

An imaging study of a phenomenon that resembles blindsight offers another perspective. In a study reported in 2006 in the *Proceedings of the National Academy of Sciences,* University College London scientists Hakwan C. Lau and Richard E. Passingham produced experimental conditions that caused variations in the conscious perception of test subjects. The paper, "Relative Blindsight in Normal Observers and the Neural Correlate of Visual Consciousness," showed how blindsight could be mimicked in unimpaired subjects. The procedure involved presenting a masking stimulus that quickly followed the presentation of a visual target stimulus the subjects were meant to observe. Immediately after the target presentation the subjects had to make a guess as to what the target stimulus was, and the subjects also reported whether or not they had consciously perceived this stimulus. Because of the mask, the subjects did not report seeing the target stimulus as often as they made correct guesses. In other words, the subjects did better than they should have if their only source of information was conscious perception—this is similar to the blindsight phenomenon. The researchers used fMRI to image the subjects during the experiment and discovered an association between the level of conscious perception and activity in certain areas of the prefrontal cortex. According to this experiment, the prefrontal cortex could be heavily involved in the formation of a conscious perception.

Other, more common phenomena are easier to study due to an increased number of patients and less difficulty in demonstrating the effect. An unfortunately prevalent disorder called schizophrenia, which strikes about 1 percent of the population, exemplifies another disconnection between perception and reality. Schizophrenia patients suffer distorted perceptions such as delusions and hallucinations, and often display confused emotions, such as laughing or crying at inappropriate moments. Imaging experiments performed by Joseph H. Callicott of the National Institutes of Health and his colleagues showed abnormal activation patterns in the prefrontal cortex of schizophrenic patients as they engaged in mental tasks requiring memory (retention of given information for a certain period of time). Other researchers have also found unusual activities in areas of a schizophrenic patient's prefrontal cortex—sometimes the activation is more than that which typically occurs for the control group (who are not patients), and sometimes less, depending on the task.

An alternative hypothesis of the nature of consciousness focuses on the specific properties of the activities across multiple regions of the brain. If consciousness is not somehow orchestrated by a single area (as proposed by the hypothesis above), then the link between areas probably comes about because of some feature of the activities in each region. One feature that neuroscientists have started investigating is the timing of the activity. The idea is that the multiple regions underlying the perception of a particular object, such as a red car traveling down the street, may link with one another because their activity patterns are related. Perhaps consciousness and perception correspond to moments when the activity of the different regions occurs at the same time or in some kind of sequential order. The brain has long been known to engage in activity patterns such as oscillations, as described in the following sidebar.

Oscillations and other timed patterns may be relevant to brain functioning, but proving this is true is not an easy task. Rhythms have been recorded in EEGs ever since Hans Berger's experiments in the 1920s, and the waves tend to occur during specific states of arousal, but the rhythms may be a byproduct of the brain's functioning rather than the player of a vital role in producing those functions. While the EEG offers excellent time resolution so that it can record even high-frequency oscillations, the recordings combine the activities of many areas of the brain, and such poor spatial resolution does not let scientists probe single networks and relate their activities to each other. The metabolic imaging techniques of fMRI and PET offer good spatial resolution but do not generally have sufficient time resolution to record oscillations.

Many experiments using the single neuron recording technique in animals have suggested a role for gamma waves, particularly those around 40 hertz, in the processing of sensory information. One of the pioneering studies appeared in *Nature* in 1989 ("Oscillatory Responses in Cat Visual Cortex Exhibit Inter-columnar Synchronization which Reflects Global Stimulus Properties"), when Charles M. Gray, Peter

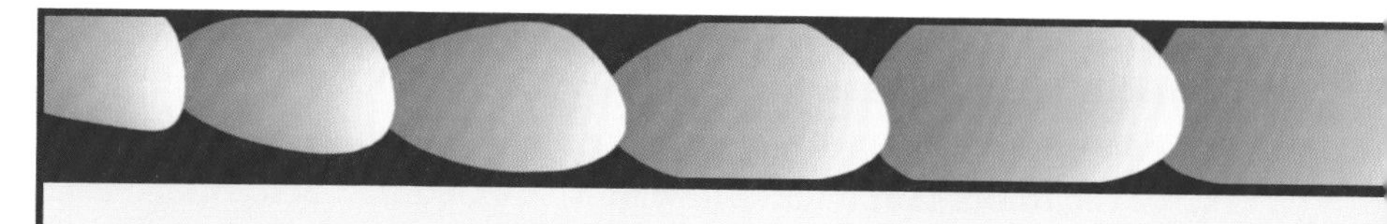

Neural Oscillations

Brain rhythms or oscillations have been recorded since the development of the EEG. The recorded activity varies over time, and sometimes the activity varies rhythmically so that the line traced by the machine begins to oscillate, going up and down at a certain frequency. The pattern resembles a wave. Hans Berger discovered one of the first waves, whose frequency ranged between eight and 13 cycles of the wave per second. This oscillation became known as Berger's wave or Berger's rhythm. (The unit of frequency is known as the hertz, named after German physicist Heinrich Hertz [1857–94], and equals one cycle per second. The frequency of Berger's wave is therefore 8–13 hertz.)

Researchers subsequently found waves of many frequencies in EEGs, and grouped the waves according to frequency, naming the groups with letters of the Greek alphabet. Berger's waves became known as alpha waves. Beta waves are roughly in the 14–30 hertz range, gamma waves 30–70 hertz, delta waves 1–4 hertz, and theta waves 4–8 hertz.

Although scientists do not yet know exactly how and why the brain generates these activity patterns, the rhythms tend to correspond to mental states. Alpha waves are associated with rest—if a person closes his or her eyes during an EEG recording (but remains awake), alpha waves will usually appear a short time later, particularly on the electrodes pasted at the back of the head. A person who is busy thinking about something will generate beta and gamma waves, a drowsy person generates theta waves, and delta waves often occur during sleep.

König, Andreas K. Engel, and Wolf Singer at the Max Planck Institute for Brain Research in Frankfurt am Main, Germany, found synchronized oscillations in a cat's visual cortex. The oscillations involved different aspects of the same visual stimulus, and synchronization could possibly be the solution to the mystery of how different parts of an object become linked into a single perception.

Synchronization might prove to be important in perception, but it cannot last long or involve many neurons, for when too many neurons are active at the same time, a seizure results. Since the oscillations that may be involved in perception and information processing are brief and have small amplitudes, they are not easy to study even with an imaging technique such as MEG, which has excellent time and spatial resolution. But intrepid scientists are making the attempt. In 2005, for example, Bernhard Ross and Anthony T. Herdman at the Rotman Research Institute in Toronto, Canada, and Christo Pantev at the Münster University Hospital in Munich, Germany, recorded 40 hertz oscillations with MEG as subjects listened to auditory stimuli. The researchers introduced another stimulus and observed a change in the oscillations. This change would be expected if the brain was using the oscillations to process information, since the new stimulus changed the subjects' perception.

The scientific study of complex phenomena such as perception and consciousness is still in its infancy. Researchers do not even agree on a precise definition for consciousness. Some neuroscientists do not believe a precise definition is necessary or even possible at this early stage; perhaps an adequate definition of consciousness will come only when scientists have a better understanding of its neural mechanisms.

Everyone has a notion of what consciousness is because everyone experiences his or her own mind, but the brain continues to keep some of the details of its operations a secret. Although consciousness is a result of brain activity, much research needs to be done before neuroscientists understand which activity is critical, and where in the brain it occurs. There might be a single brain region that is primarily responsible for a person's awareness, or broad expanses might be necessary, linked by oscillations or synchrony (or some combination of both). Or there could be another mechanism, as yet unknown, that neuroscientists will discover as they continue to study and image the human brain.

CONCLUSION

Human brain imaging gives neuroscientists the opportunity to peer inside the skull and study the living brain. Imaging techniques supplement and have greatly expanded earlier methods such as animal experiments, recording noisy signals from a person's scalp, and studying the deficits exhibited by brain-injured patients. With the new images neuroscientists have mapped the localization of function in the human brain, and started to make inroads on difficult topics such as learning, perception, and consciousness.

Further progress will rely on the skills with which scientists can employ imaging techniques. As imaging technology advances, clearer, higher resolution images will emerge. Activities other than cell metabolism also need to be studied. Scientists have already started to image the distribution and activity of important molecules such as receptor proteins, which play important roles in neural communication, and images of the location of these molecules tell neuroscientists what kind of communication is taking place at specific sites in the brain.

PET and fMRI produce indirect measurements of the brain's electrical activity by imaging metabolic activity. Nikos Logothetis is studying ways to create imaging techniques that are more direct reflections of electrical activity—a more accurate "mirror," as envisioned by Hans Berger in 1924. A molecule or probe that changes its magnetic properties in conjunction with a neuron's electrical activity, perhaps due to the changes in ion concentrations, could produce a signal strong enough for magnetic resonance imaging to detect. These probes are called "smart probes" because they are not rigidly fixed, but instead change their properties to indicate the presence of neural activity. Smart probes, if and when they are developed, would enhance the spatiotemporal resolution—resolution of time and space—of imaging and, as Logothetis wrote on his Web site, "their successful application in neuroscience is likely to usher in a real revolution, as it promises truly spectacular spatiotemporal resolution and specificity for whole-brain imaging."

Other exciting developments involve the use of electromagnetism not only to detect neural activity, but to influence it as well. Tiny electrical currents in the brain generate small magnetic fields that can be measured outside of the skull by sensitive instruments as in MEG. According to the laws of physics, fluctuating magnetic fields can also induce electrical currents. This means that applying strong magnetic fields to a person's head will produce currents in neurons, a painless procedure that has been performed in several laboratories. These currents tend to disrupt the activity of neurons. When researchers project the applied magnetic field to a small area of the brain, the induced currents scramble the neurons' communications and cause a temporary loss of function in that area.

This new technique, known as transcranial magnetic stimulation (TMS), can induce harmless and temporary "lesions" in humans, allowing neuroscientists to study the function of certain areas by the deficits created by the "lesion." TMS can also recreate important phenomena such as blindsight. Jennifer L. Boyer, Stephenie Harrison, and Tony Ro at Rice University reported in 2005 an experiment to produce blindsight under rigorously testable laboratory conditions. The paper, published in the *Proceedings of the National Academy of Sciences* and titled "Unconscious Processing of Orientation and Color without Primary Visual Cortex," disabled volunteers' V1 cortex. Volunteers could still react to visual stimuli under certain conditions, despite being unable to see.

These new tools increase the capacity of neuroscientists to study one of the most complicated and fascinating objects in the universe—the human brain. Neuroscience remains a developing field of biology, and no one has yet found a perfect "mirror" to reflect what is happening inside the head as people talk, think, dream, compose, and create. But progress has been rapid in the last few decades, and the future looks promising.

CHRONOLOGY

18th century B.C.E. Egyptian papyrus mentioning brain surgery is written. This is the first record concerning the nervous system.

fifth century B.C.E. Greek physician Hippocrates (ca. 460–377 B.C.E.) writes about epilepsy as a disease of the brain.

fourth century B.C.E. Greek philosopher Aristotle (384–322 B.C.E.) proposes that the heart is the source of human intelligence.

second century C.E. Greek physician Galen (129–99), living in the Roman Empire, gives lectures on the brain's anatomy.

1791	Italian physician Luigi Galvani (1737–98) publishes his work on electricity and the contractions of frog legs.
1808	Austrian anatomist Franz Joseph Gall (1758–1828) publishes early phrenology studies.
1839	German biologist Theodor Schwann (1810–82) proposes the cell theory—all organisms are composed of cells.
1848	An iron spike pierces the frontal lobe of railroad worker Phineas Gage (1823–60) in Vermont, causing noticeable personality changes.
1854	French physician Louis Pierre Gratiolet (1815–65) names the lobes of the brain.
1861	French surgeon and scientist Paul Broca (1824–80) announces a theory that language production is localized in a specific area, which became known as Broca's area, in the left hemisphere of the brain.
1870	German biologists Eduard Hitzig (1838–1907) and Gustav Fritsch (1838–1927) discover motor cortex in dogs.
1875	British scientist Richard Caton (1842–1926) records electrical activity from electrodes attached to the surface of the brain of experimental animals.
1889	Spanish anatomist Santiago Ramón y Cajal (1852–1934) argues that the brain consists of separate, communicating cells rather than a mesh of continuous elements.
1890	British scientist Sir Charles Scott Sherrington (1857–1952) and Scottish scientist Charles Smart Roy (1854–97) publish their findings of blood flow regulation in the brain.

1891	German scientist Wilhelm von Waldeyer-Hartz (1836–1921) coins the term *neuron.*
1897	Sir Charles Scott Sherrington names and describes the synapse.
1929	German psychiatrist Hans Berger (1873–1941) begins to publish his work with the EEG.
1950s	David Hubel and Torsten Wiesel begin recording from single neurons in the thalamus and cortex of the cat's visual system.
1973	Michael M. Ter-Pogossian, Edward Hoffman, and Michael E. Phelps develop the first PET scanner for human imaging at Washington University in St. Louis, Missouri.
	Paul Lauterbur, building on the work of numerous other researchers, produces the first magnetic resonance image of an animal (a clam).
1980s	Marcus Raichle, building on the work of numerous other researchers, pioneers the development of PET to study human brain function.
1990	United States president George Herbert Walker Bush pronounces the 1990s as the Decade of the Brain, calling attention to the rapidly advancing discipline of neuroscience.
1991	Jack Belliveau and colleagues publish the first report that magnetic resonance imaging can be applied to the functioning brain, initiating the fMRI technique.
1994	Scientists meet in Tucson, Arizona, for an international conference, Toward a Scientific Basis of Consciousness.

2005	Rice University researchers Jennifer L. Boyer, Stephenie Harrison, and Tony Ro use transcranial magnetic stimulation to produce blindsight in subjects under laboratory conditions.
2006	University College London scientists Hakwan C. Lau and Richard E. Passingham develop an experiment to mimic the phenomenon of blindsight in unimpaired subjects.
2009	The 15th international scientific conference on consciousness, Toward a Science of Consciousness, takes place in Hong Kong.

FURTHER RESOURCES

Print and Internet

Binder, Jeffrey R., Julie A. Frost, Thomas A. Hammeke, Robert W. Cox, et al. "Human Brain Language Areas Identified by Functional Magnetic Resonance Imaging." *Journal of Neuroscience* 17 (January 1, 1997): 353–362. Binder and his colleagues used fMRI to image the brains of 30 subjects while the subjects listened to spoken words.

Boyer, Jennifer L., Stephenie Harrison, and Tony Ro. "Unconscious Processing of Orientation and Color without Primary Visual Cortex." *Proceedings of the National Academy of Sciences* 102 (November 15, 2005): 16,875–16,879. These researchers disabled volunteers' V1 cortex, but the volunteers could still react to visual stimuli under certain conditions, despite being unable to see.

Bush, George H. W. "Presidential Proclamation 6158." July 17, 1990. Available online. URL: http://www.loc.gov/loc/brain/proclaim.html. Accessed March 30, 2009. President George H. W. Bush proclaims the 1990s as the Decade of the Brain.

Chudler, Eric H. "Milestones in Neuroscience Research." Available online. URL: http://faculty.washington.edu/chudler/hist.html. Accessed April 1, 2009. Chudler, a professor at the University of

Washington, maintains this detailed chronology of important events in the history of neuroscience.

Crick, Francis C., and Christof Koch. "What Is the Function of the Claustrum?" *Philosophical Transactions of the Royal Society B* 360 (2005): 1,271–1,279. Crick and Koch describe this as yet little known area, which lies hidden beneath the cerebral cortex at the side of the head, and its possible involvement in consciousness.

Davis, Karen D. *New Techniques for Examining the Brain.* New York: Chelsea House Publishers, 2007. Written for young adults, this book explains brain-imaging techniques in detail, and describes how the results have been used in science and medicine.

Devlin, Hannah. "Introduction to fMRI." Available online. URL: http://www.fmrib.ox.ac.uk/resources/education/fmri/introduction-to-fmri/. Accessed April 1, 2009. Hosted by Oxford University, these pages explain the theory and practice of fMRI.

Finger, Stanley. *Origins of Neuroscience: A History of Explorations into Brain Function.* Oxford: Oxford University Press, 2001 (reprint). This comprehensive and well-illustrated book follows the sometimes winding but always interesting path scientists have taken in studying the brain.

Fleischman, John. *Phineas Gage: A Gruesome but True Story about Brain Science.* Boston: Houghton Mifflin, 2002. Phineas Gage was a railroad worker who survived a horrific accident in 1848 that drove an iron rod through the front part of his brain. Although Gage recovered physically, his personality changed due to the damage to his frontal lobe, as described in this book written for young adult readers.

Gray, Charles M., Peter König, Andreas K. Engel, and Wolf Singer. "Oscillatory Responses in Cat Visual Cortex Exhibit Inter-columnar Synchronization which Reflects Global Stimulus Properties." *Nature* 338 (March 23, 1989): 334–337. The researchers found synchronized oscillations in a cat's visual cortex.

Greenfield, Susan A. *The Human Brain: A Guided Tour.* New York: Basic Books, 1998. The author, a noted British neuroscientist, gave the Royal Institution Christmas Lectures in 1994. This book is based on those talks, in which Greenfield reviews the basic principles of the brain and discusses topics such as development, vision, drug abuse, and memory in a lively and entertaining style.

Hubel, David. "Eye, Brain, and Vision." Available online. URL: http://hubel.med.harvard.edu/. Accessed April 1, 2009. Nobel laureate David Hubel offers his book, *Eye, Brain, and Vision,* revised in 1995, on the Web. The book explores Hubel's research on the visual system, for which he shared the 1981 Nobel prize in physiology or medicine with his colleague Torsten Wiesel, and explains the principles of how the eye and brain work.

Johnson, Keith A., and J. Alex Becker. "The Whole Brain Atlas." Available online. URL: http://www.med.harvard.edu/AANLIB/home.html. Accessed April 1, 2009. This Web site provides a fantastic set of images of healthy and diseased brains.

Lau, Hakwan C., and Richard E. Passingham. "Relative Blindsight in Normal Observers and the Neural Correlate of Visual Consciousness." *Proceedings of the National Academy of Sciences* 103 (December 5, 2006): 18,763–18,768. The researchers showed how blindsight can be mimicked in unimpaired subjects.

LeDoux, Joseph. *The Synaptic Self: How Our Brains Become Who We Are.* New York: Penguin, 2003. LeDoux, a professor at New York University's Center for Neural Sciences, emphasizes the importance of the pathways by which information flows in the brain—pathways created by neurons communicating via synapses. The author describes his view that these pathways are what makes people who they are.

Logothetis, N. K., J. Pauls, M. Augath, T. Trinath, and A. Oeltermann. "Neurophysiological Investigation of the Basis of the fMRI Signal." *Nature* 412 (July 12, 2001): 150–157. In a careful series of experiments using a specially designed magnet, Logothetis and his colleagues report a relationship between electrical activity and the fMRI image of an experimental animal.

Public Broadcasting Service. "The Secret Life of the Brain." Available online. URL: http://www.pbs.org/wnet/brain/index.html. Accessed April 1, 2009. This Web site is the companion online presence of the television series, which aired on PBS in 2002. The site includes a great deal of information on brain imaging.

Raichle, Marcus E. "Cognitive Neuroscience: Bold Insights." *Nature* 412 (July 12, 2001): 128–130. Raichle, a pioneer in the use of imaging tools in neuroscience, reviews the paper of Logothetis and his colleagues, "Neurophysiological Investigation of the Basis of the fMRI Signal," which appears in the same issue of *Nature.*

Shapiro, Kevin A., Lauren R. Moo, and Alfonso Caramazza. "Cortical Signatures of Noun and Verb Production." *Proceedings of the National Academy of Sciences* 103 (January 31, 2006): 1,644–1,649. These researchers took fMRI images of the brains of people who produced either verbs or nouns in short phrases, and discovered that two areas of the brain were activated more strongly during verb production.

Web Sites

Laboratory of Neuro Imaging at the University of California, Los Angeles. Available online. URL: http://www.loni.ucla.edu/. Accessed April 1, 2009. The home page of this laboratory describes its equipment and research projects, provides examples of images, and discusses highlights from recent findings.

Logothetis, Nikos K. Available online. URL: http://www.kyb.mpg.de/~nikos. Accessed April 1, 2009. Nikos Logothetis, a professor at the Max Planck Institute for Biological Cybernetics, discusses his research in an understandable and informative style.

Salk Institute for Biological Sciences. Available online. URL: http://www.salk.edu/. Accessed April 1, 2009. The home page for the Salk Institute provides news and information on its latest research projects.

Society for Neuroscience. Available online. URL: http://www.sfn.org. Accessed April 1, 2009. The Society for Neuroscience is a nonprofit organization of scientists and physicians who study nervous systems. In addition to providing research information, the society's home page offers tutorials and news written for the general public.

2

THE HUMAN GENOME IN HEALTH AND DISEASE

The atomic bombs that the United States dropped on Japan in 1945 to end World War II ushered in the new and often frightening era of nuclear energy. While this form of energy is capable of producing devastating weapons, it can also be harnessed in a more controlled fashion to generate electricity. But the danger of nuclear reactions does not lie only in the heat and crippling blast of a bomb's detonation—radiation and high-energy particles associated with the reactions produce deadly effects of their own. Cells and tissues of the body are susceptible to damage from radioactivity, whether produced in small amounts by controlled nuclear reactions or all at once, as in the fallout from a nuclear explosion.

The United States government created the Atomic Energy Commission (AEC) in 1946 to study nuclear reactions and their possible applications, as well as the threats they may pose to human health and welfare. Researchers found that in a lot of cases the harmful effect of radiation arises from a mutation—a change in the body's genetic material that carries *inherited traits.* Because of the susceptibility of genetic material to radiation damage, the AEC became interested in the structure and properties of these molecules. Scientists associated with atomic physics tend to think big, despite the tiny size of the subject of their study; the Manhattan Project, which produced the first atomic bomb, exemplifies this kind of thinking, as does another early goal—to find the entire set of human genes, the units of inheritance. Knowledge of the entire set would make the study of genetics much simpler than having to attack the problem one gene at a time.

Technology to study the genome—all of the genetic material of an organism—was in its infancy during the 1940s and '50s. But the goal was not forgotten. The government abolished the AEC in 1974 but in 1977 created a new department, the Department of Energy (DOE), charged with planning and researching a variety of issues related to energy, including nuclear energy. Shortly thereafter, the DOE turned some of its attention toward the ambitious goal of examining all human DNA—the Human Genome Project. This project mapped the human genome, giving biologists and other researchers unprecedented access to the material that is the "blueprint," or plan, of a human being.

A person's genes are responsible for much of what makes each person a unique combination of talents, strengths, and weaknesses, including susceptibility to disease. Although the large amount of data contained in the human genome will require a lot of time and effort to analyze, researchers believe it will help them gain a much better understanding of how human beings work. This chapter describes genomic research that is probing DNA's role in the variability of human beings and the differences in susceptibility to disease and responses to medication.

INTRODUCTION

The identity of the molecules that carry inherited traits is a recent discovery. For a time in the early 20th century many scientists thought proteins were probably the carriers of the genetic information that passes from parent to offspring.

Gregor Mendel (1822–84), an Augustinian monk at Brünn, in Austrian Moravia (now Brno, in the Czech Republic), proposed the existence of certain factors that determine inherited traits. In the late 1850s, Mendel started breeding pea plants and using statistics to keep track of the number of plants with traits such as flower color and seed texture. Mendel found that a pair of genetic factors, which later became known as genes, controlled each trait.

But Mendel's simple ideas of inheritance did not work well with more complex organisms such as mammals. The work of Mendel was ignored for many years even though Charles Darwin's (1809–82) theory of evolution, published in 1859, relied on genetic variation, inheritance, and a process of natural selection of traits that increase an organism's chances to survive and reproduce. But by the early 20th century, people began to realize that Mendel's concept of genes could help explain how evolution occurs and how traits propagate through populations. Working with fruit flies, Thomas H. Morgan (1866–1945), a scientist at Columbia University in New York, and his colleagues found that genes reside on chromosomes.

The primary components of chromosomes are DNA and proteins. DNA is a *nucleic acid* consisting of a chain of four types of molecules known as *nucleotides,* or nucleic *bases,* connected to one another with strong chemical bonds. (The term *nucleic acid* derives from the location of DNA, most of which is in an important structure in the cell called a nucleus. A cell's nucleus should not be confused with an atom's nucleus, which is the origin of terms such as nuclear energy and nuclear reactions.) Protein is a chemically bonded chain of 20 different types of molecules called *amino acids.* Because there is a lot of genetic information contained within an organism, particularly complicated organisms such as humans, the best bet for genes seemed to be proteins, since their greater variety and complexity would seem to give them an increased capacity to store information.

But experiments proved otherwise. In 1928, Frederick Griffiths (1879–1941) discovered a component of bacteria that conveyed traits such as toxicity to other bacteria, transforming them. Griffiths did not know the chemical identity of the component he found, but in 1944 Oswald Avery (1877–1955) and his colleagues Maclyn McCarty (1911–2005) and Colin MacLeod (1909–72) succeeding in isolating this transforming agent. It proved to be DNA.

The properties of DNA that make it an ideal storage molecule for information came to light in 1953. In a short paper in *Nature* titled "Molecular Structure of Nucleic Acids," James Watson (1928–   ) and Francis Crick (1916–2004) at the Cavendish Laboratory in Cambridge, England, proposed that the structure of DNA was a double helix, as shown in the figure on page 42. Part of the inspiration for their work was a set of experiments conducted by Maurice Wilkins (1916–2004) and Rosalind Franklin (1920–58), who probed DNA with X-rays. Later experiments showed that genetic information is contained in the sequence of bases.

GENES AND DNA

The double helix structure of DNA forms when two strands join in a special way. Each strand has a "backbone" composed of sugar

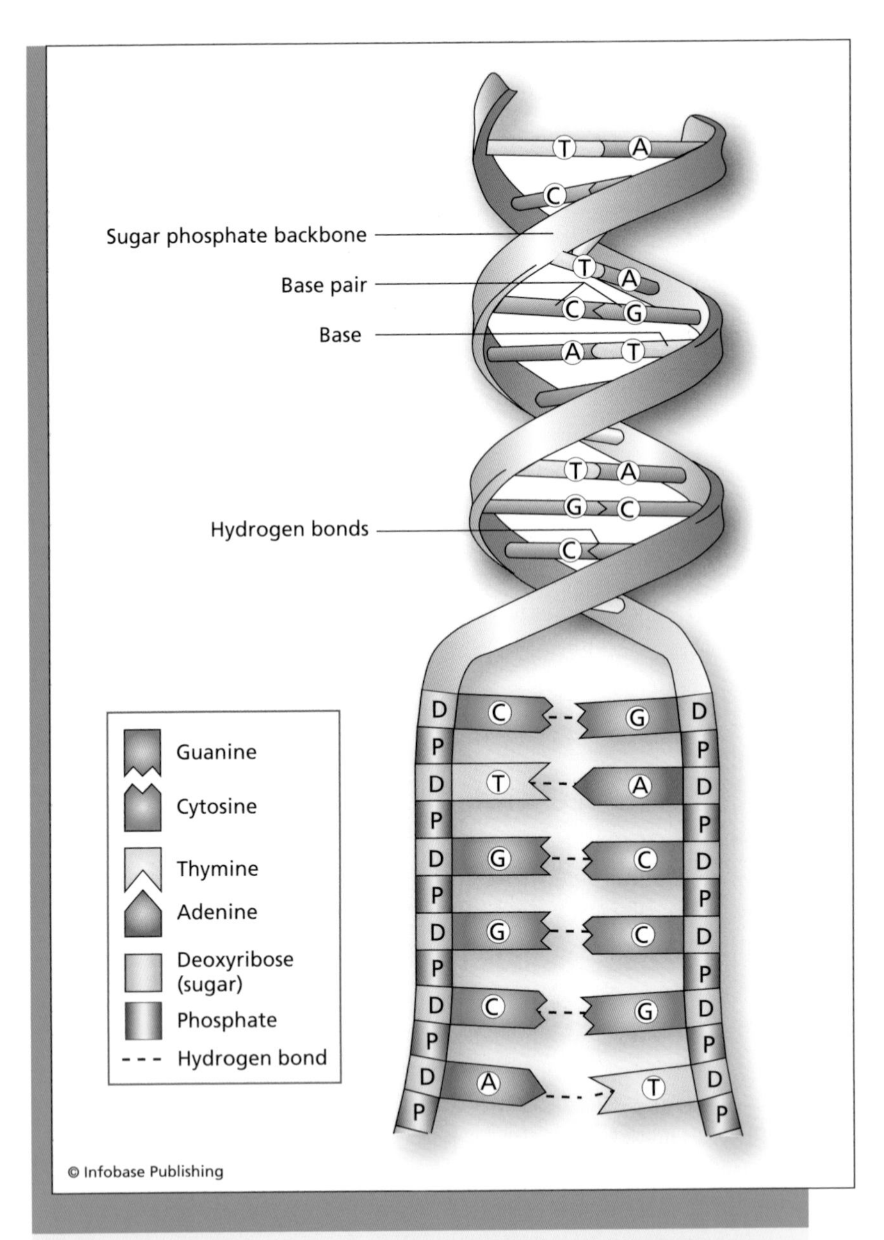

DNA is a double helix with a backbone, composed of phosphates and the sugar deoxyribose, and bases (adenine, thymine, guanine, and cytosine) connected with weak but important bonds called hydrogen bonds. Because of the structure of the bases (shown diagrammatically rather than realistically in the figure), adenine joins only with thymine and guanine joins only with cytosine.

molecules known as deoxyribose, joined by phosphate groups (containing four oxygen atoms bonded to an atom of phosphorus). Attached to the sugar molecule is a base, which may be either adenine (A), thymine (T), guanine (G), or cytosine (C). Hydrogen bonds join the bases of two strands, forming the helix. These bonds arise because when hydrogen shares its electrons to create another type of bond—a covalent bond with other atoms such as oxygen and nitrogen—the other atoms have a stronger pull on the shared electrons than hydrogen (which is the smallest atom). This unequal sharing creates a region of electric charge. The charge is positive around covalently bonded hydrogen atoms and negative around oxygen and nitrogen atoms, so the hydrogen atom is electrically attracted to the oxygen and nitrogen, including oxygen and nitrogen atoms engaged in similar covalent bonds in other molecules. This attraction is the basis for hydrogen bonds.

Hydrogen bonds are about 20 times weaker than covalent bonds, but they are strong enough to hold together two strands of DNA and keep the molecule stable. Due to the size and shape of the bases, A normally forms stable hydrogen bonds only with T, and C forms stable hydrogen bonds only with G. A and T are said to be complementary, as are C and G. The hydrogen bonds hold complementary strands together.

The sequence of bases contains the coded information of *genes.* This sequence provides the necessary information to construct a protein, or in some cases, a molecule of *ribonucleic acid* (RNA). (Some people use the term *gene* to refer only to segments of DNA that code for proteins, but there is no universally accepted definition of this term. In this chapter, gene refers to a DNA segment that codes either a protein or an RNA molecule, or, in general, a unit of inheritance.) Proteins, as mentioned earlier, are strings of amino acids, and RNA molecules are strings of nucleic bases, similar to DNA except that the sugar is ribose rather than deoxyribose, uracil (U) replaces thymine, and RNA does not generally form a double helix structure. For proteins, three DNA bases comprise a *codon,* which codes for an amino acid. Each base of an RNA molecule is encoded by a single DNA base.

Proteins perform a number of critical functions in cells and tissues. As discussed in chapter 3, the sequence of amino acids determines a protein's shape, which in turn governs what functions the protein will serve. Many proteins act as enzymes, speeding up chemical reactions (without enzymes, many chemical reactions that are essential for life would occur too slowly), while other proteins play a role in structural support, transportation of nutrients, movement, or other functions. With 20 "letters" (amino acids) in a protein's "alphabet," these molecules offer a lot of variety. This complexity explains why many scientists initially assumed proteins were the storage units of genetic information. But the mechanism of storage proved to be simpler, and is shared by all living organisms on Earth. Virtually every organism uses exactly the same genetic code, with the same codons specifying the same amino acid. There are 64 possible codons—the number of ways to arrange three letters, each having four possibilities, is $4 \times 4 \times 4 = 64$—as listed in the following table, using the RNA bases.

RNA also plays a number of roles in the body, particularly concerning DNA. One of RNA's primary roles is to help replicate DNA, such as when a cell divides and needs to copy its DNA for each daughter cell, and to "read" the genetic code. In order to access the information, special enzymes help the DNA helix to unzip, breaking the weak hydrogen bonds. In the process of reading the code, an association of protein and RNA molecules copies a DNA sequence onto a new RNA molecule. In other words, the DNA is a template in this process, which is called *transcription.* The complementary base pairing is critical, as suspected ever since DNA's structure had been found; in the 1953 *Nature* paper of Crick and Watson, they write with considerable understatement, "It has not escaped our notice that the specific pairing we have postulated immediately suggests a possible copying mechanism for the genetic material."

As shown in part (a) of the figure on page 46, the enzymes transcribe one strand by matching each DNA base with its complement, though the match is made with RNA bases, with uracil substituting for thymine. Then the hydrogen bonds form again, zipping the helix back together. If the end product is an RNA molecule, the job is over. For proteins, the transcribed RNA is known as *mRNA,* short for messenger RNA. This RNA carries the instructions to make the specific protein, as copied from the DNA template, to another protein-RNA enzyme complex that catalyzes the bonding of amino acids in correct order to make the protein, a process called *translation.* As each codon is read, the enzyme complex inserts the appropriate amino acid into the growing chain. When the stop signal is reached, the process terminates, and the new, correctly sequenced protein can fulfill its function.

THE GENETIC CODE	
Amino Acid	**Codons**
Alanine	GCU, GCC, GCA, GCG
Arginine	CGU, CGC, CGA, CGG, AGA, AGG
Asparagine	AAU, AAC
Aspartate	GAU, GAC
Cysteine	UGU, UGC
Glutamine	CAA, CAG
Glutamate	GAA, GAG
Glycine	GGU, GGC, GGA, GGG
Histidine	CAU, CAC
Isoleucine	AUU, AUC, AUA
Leucine	UUA, UUG, CUU, CUC, CUA, CUG
Lysine	AAA, AAG
Methionine (or start codon, the point at which the protein begins)	AUG
Phenylalanine	UUU, UUC
Proline	CCU, CCC, CCA, CCG
Serine	AGU, AGC, UCU, UCC, UCA, UCG
Threonine	ACU, ACC, ACA, ACG
Tryptophan	UGG
Tyrosine	UAU, UAC
Valine	GUU, GUC, GUA, GUG
Stop codon (the point at which the protein ends)	UAA, UAG, UGA

Human cells have a total of about three billion bases contained in 23 pairs of chromosomes. There are two copies of each of 22 different

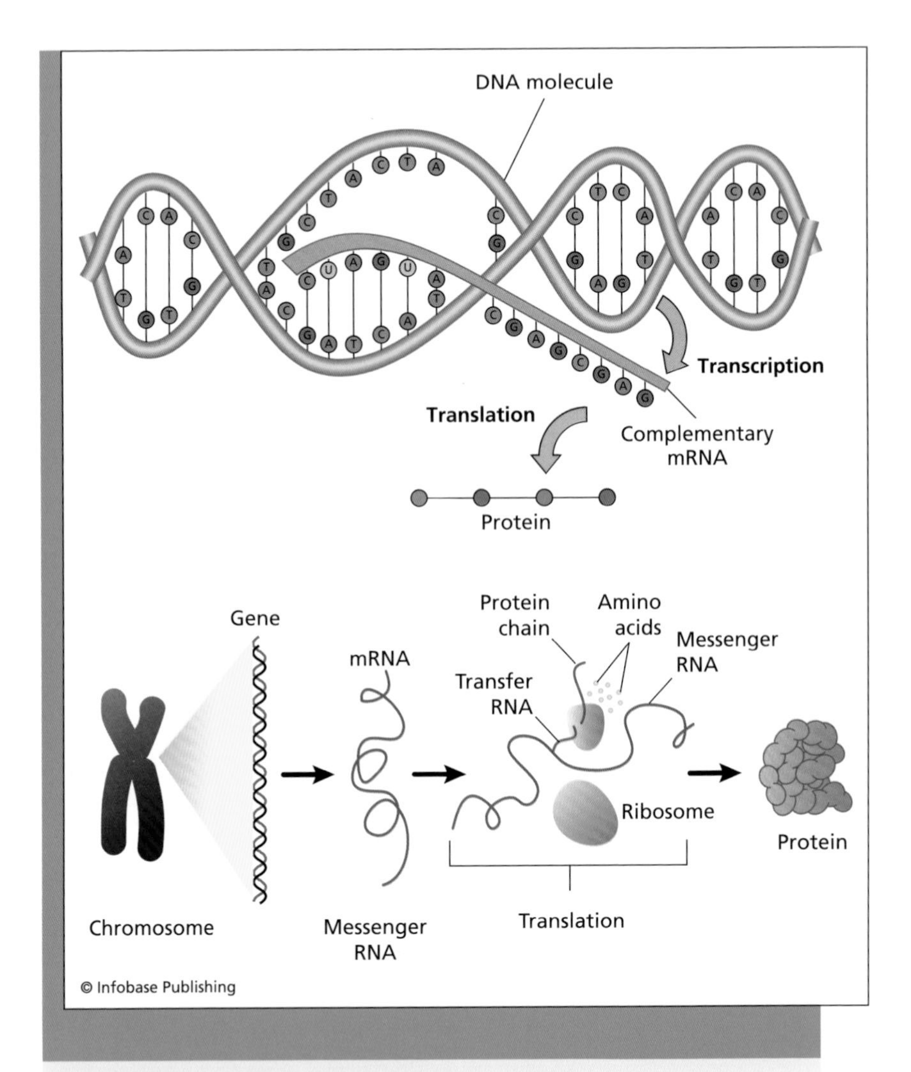

(A) An mRNA copy of the DNA sequence is made during transcription.
(B) To make a protein, the mRNA is translated. Enzymes of the ribosome read each codon and link the appropriate amino acid to the growing chain.

chromosomes, known as autosomes, and one pair of sex chromosomes. Each chromosome contains some number of genes, and having a pair of each chromosome means that cells have two copies of every gene. (An exception occurs in males, whose sex chromosomes, X and Y, are different.) But although these two gene copies code for the same product—a specific protein or RNA—the exact sequence may not be the same. Some variation exists, and the differences in the two sequences may result in proteins with slightly different shapes. (Sometimes only one of the two copies of a gene in an individual is used, and the other is ignored.) These different gene versions are known as *alleles.*

Some proteins do a better job under certain conditions than others, giving an advantage to individuals who have the corresponding allele. Evolution selects the fittest individuals, who thrive and reproduce and transmit their advantageous alleles (genes) to their offspring.

Not all cells require all proteins. For example, cells in the brain have proteins that help them send electrochemical messages to one another, but a cell in the liver has no need of these proteins. Cells also may need certain proteins only at specific times, such as during division. Transcribing genes and making proteins requires energy, and cells reduce waste by regulating these processes. Genes are expressed—the gene product is made—only in certain cells and at certain times. Molecules that bind to specific DNA sequences regulate gene expression.

Although cells minimize waste by regulating gene expression, a striking feature of DNA in humans is that much of it does not seem to code for either RNA or proteins. For instance, in many organisms, mRNAs contain sequences that are cut out by special enzymes before the protein is made. A deleted sequence is called an *intron,* whereas the remaining sequences, which are used to make the protein, are called *exons.* (Introns refer to intervening sequences, while exons are expressed.) In addition to introns, chromosomal DNA in humans and most other multicellular organisms include vast regions outside of genes that are not used for coding. Much of these regions contain noncoding sequences that repeat numerous times. But while this repetitive DNA may not make RNA or protein, some of it has important applications in the technology of DNA forensics, as described in the following sidebar.

Half of a person's DNA comes from the father and half from the mother, but every person has a unique mixture of genes and sequences. (The only exceptions are identical twins.) A person's DNA not only provides a "fingerprint" for identification purposes, it also bestows upon each individual a unique set of traits and tendencies, as well as susceptibilities to certain diseases.

Not all of a person's traits are due to genetics—environmental factors, upbringing, personal choices, and so forth influence many vital characteristics—but genes are extremely important. Genetics, the study

DNA Forensics

Identical twins develop from the same fertilized egg cell (the union of egg and sperm), and therefore share the same set of genes. (Most twins do not develop from the same egg cell, and do not have an identical set of genes. These twins are known as fraternal twins.) No one except pairs of identical twins share the same full set of DNA sequences, although the vast majority of DNA from all humans is the same, especially among blood relatives, who inherit similar genetic material. One person's DNA differs from another person's in only about one out of every 1,000 bases on average. Although this is not a high percentage, the differences are enough to distinguish one person from another (except identical twins), similar to identification mechanisms based on fingerprints, which can distinguish even identical twins. (Identical twins do not have identical fingerprints because the development of fingerprints is not due solely to genetic factors.) Identification based on DNA is often called DNA fingerprinting.

Sir Alec Jeffreys (1950–), a scientist at the University of Leicester in the United Kingdom, pioneered DNA fingerprinting. In 1985, Jeffreys used the technique to prove the parentage of a boy who had been prevented from entering Britain because authorities suspected his passport had been forged. The first use of the technology in a criminal case occurred the following year, when Jeffreys showed that a man who had confessed to a murder and rape in Leicester could not have been guilty because his DNA did not match the crime scene DNA. After proving the confession was false,

of inherited traits, has been an active scientific subject especially since Mendel's time, but until recently scientists have studied genes and other DNA sequences mostly one at a time, singled out from the whole. Rethe police took DNA samples from local men, one of whom proved to be a match.

But even today, DNA technology is not fast enough to sequence a person's entire DNA in a reasonable period of time and for a reasonable amount of money. Shortcuts had to be found. Early DNA fingerprinting methods made use of restriction fragment length polymorphism (RFLP). Special enzymes called restriction enzymes cut DNA only at specific sequences, resulting in a set of fragments of various lengths. For instance, a piece of DNA may have nine of these specific sequences, so an enzyme will cut it nine times. DNA from another person is slightly different and may have only eight of these specific sequences, so the fragments from this sample will be slightly different—the term *polymorphism* means having a different shape or form.

Most methods today make use of repetitive DNA known as tandem repeats because their number is highly variable among individuals. Some people may have 33 repeats at a specific site, other people only 19. In the United States, the Federal Bureau of Investigation uses a system called CODIS (Combined DNA Index System), in which tandem repeats at 13 sites are counted. The chances of two people having the same number of repeats at all 13 sites depends on their genetic relatedness, but typically the odds are one in millions or even billions.

DNA technology can also be used in resolving old mysteries and exonerating people who have been falsely convicted. The Innocence Project, an organization dedicated to investigating such cases, has used DNA testing to exonerate 234 prisoners as of April 1, 2009.

searchers sequenced bits and pieces of human DNA, acquiring knowledge of the function of those bits and pieces but always failing to grasp the "big picture."

HUMAN GENOME PROJECT

The decision to undertake the ambitious project of sequencing the human genome was not an easy one. Everyone understood the benefits, but some people, including scientists, questioned whether the project was worth the time and expense. In the 1980s, when the DOE began a serious consideration of the project, the technology to sequence DNA was slow and laborious, requiring weeks or months for even a small sample. Some people were worried that the project would require so much time and money that it would drain all resources from other worthy DNA studies.

But genetics researchers were spending a huge portion of their time trying to find genes associated with certain traits. Researchers had found and sequenced only a few thousand genes by the 1980s, which was only a small portion of the total. Years of effort were usually required to track down a single gene; for example, Nancy Wexler and her colleagues spent more than a decade finding the gene that causes a disease known as Huntington's Chorea (the researchers finally succeeded in 1993). If the full sequence of the genome were made available, a lot of the work involved in these searches would already be done. And researchers could learn much more by analyzing the entire sequence instead of just bits and pieces of it.

Scientists decided that the advantages of sequencing the genome justified the time and expense of the project, and on October 1, 1990, DOE and the National Institutes of Health (NIH) launched the Human Genome Project. NIH, the subject of the following sidebar, is the main funding agency for biomedical research in the United States. The schedule called for completion in 15 years. Primary goals included sequencing all three billion bases of the human genome, locating and identifying all genes, and freely distributing this information to the public and to researchers across the globe. Making the data public was a vital part of the plan—the project's mission was not just to help scientists at DOE and NIH, but also to assist the research of all scientists. The end result would be an acceleration of genetic discoveries to the benefit of humans worldwide.

The workload would be shared among researchers at various locations. Since the process of sequencing DNA and analyzing the results was a time-consuming task, DOE and NIH funded sequencing efforts at a large number of universities and laboratories across the United States.

Automated machines find the sequence of DNA samples at the Center for Genome Research of the Whitehead Institute at the Massachusetts Institute of Technology *(Sam Ogden/Photo Researchers, Inc.)*

At any given time during the Human Genome Project, the program provided funds to about 200 different researchers.

Although the project aimed to sequence the genome of human beings, in reality there is no such thing as "the" genome of humans or any other species. What exists are large numbers of individuals who possess

National Institutes of Health (NIH)

The National Institutes of Health, usually known by the acronym NIH, conducts and sponsors research aimed at advancing the biological sciences and improving medicine. The plural—"institutes"—is because NIH is composed of several branches, or institutes, devoted to specific areas of research. There is the National Cancer Institute, National Institute on Drug Abuse, National Human Genome Research Institute, and 24 other institutes or centers.

NIH had a humble beginning. In 1887, the Marine Hospital Service founded a single-room laboratory on Staten Island, New York, to study infectious diseases. This laboratory became known as a laboratory of hygiene, which refers to the science of health and prevention of disease. The Hygienic Laboratory grew and on May 26, 1930, became the National Institute of Health. Five years later, Mr. and Mrs. Luke I. Wilson donated their estate in Bethesda, Maryland, for the institute's use. Bethesda became the institute's headquarters in 1938, and remains so today. The number of branches grew, and the National Institute of Health became the National Institutes of Health in 1948.

The budget grew accordingly. Funds for NIH come from taxpayers and are appropriated by Congress. NIH's total budget for 2008 exceeded $29 billion. About 20 percent of this money supports scientists who work in NIH laboratories, and most of the rest is distributed to researchers working at various universities and laboratories across the country. Funding is highly competitive. Researchers who wish to receive support, usually called a research grant, must write an extensive proposal outlining their goals and the steps they plan to take to achieve them. For instance, a scientist at a university may wish to study some particular issue in the genetics of cancer. He or she would do some preliminary research, and then write a detailed proposal specifying the questions to be answered and why they are important, and

View of the National Institutes of Health (NIH) campus in Bethesda, Maryland *(NIH)*

discuss the experiments or theoretical models necessary to find the solutions. NIH receives thousands of grant applications every year, but the budget allows only about a quarter or a third of these applications to be approved. Expert scientists, selected by NIH, convene panels and debate the merits of each proposal in their field of expertise, choosing what they consider the best ones for funding. (Panelists are not allowed to debate their own research grants, of course!)

Money for special projects such as the Human Genome Project is also available. In addition, NIH sponsors training grants, given to educational institutions such as universities to train future scientists. People who wish to become scientists typically attend graduate school after having received a bachelor's degree in college. Many people who pursue a career in research obtain the terminal research degree in graduate school (*terminal* in the sense of final, or highest degree)—a doctor of philosophy, abbreviated Ph.D. (from the Latin expression *philosophiae doctor,* meaning teacher).

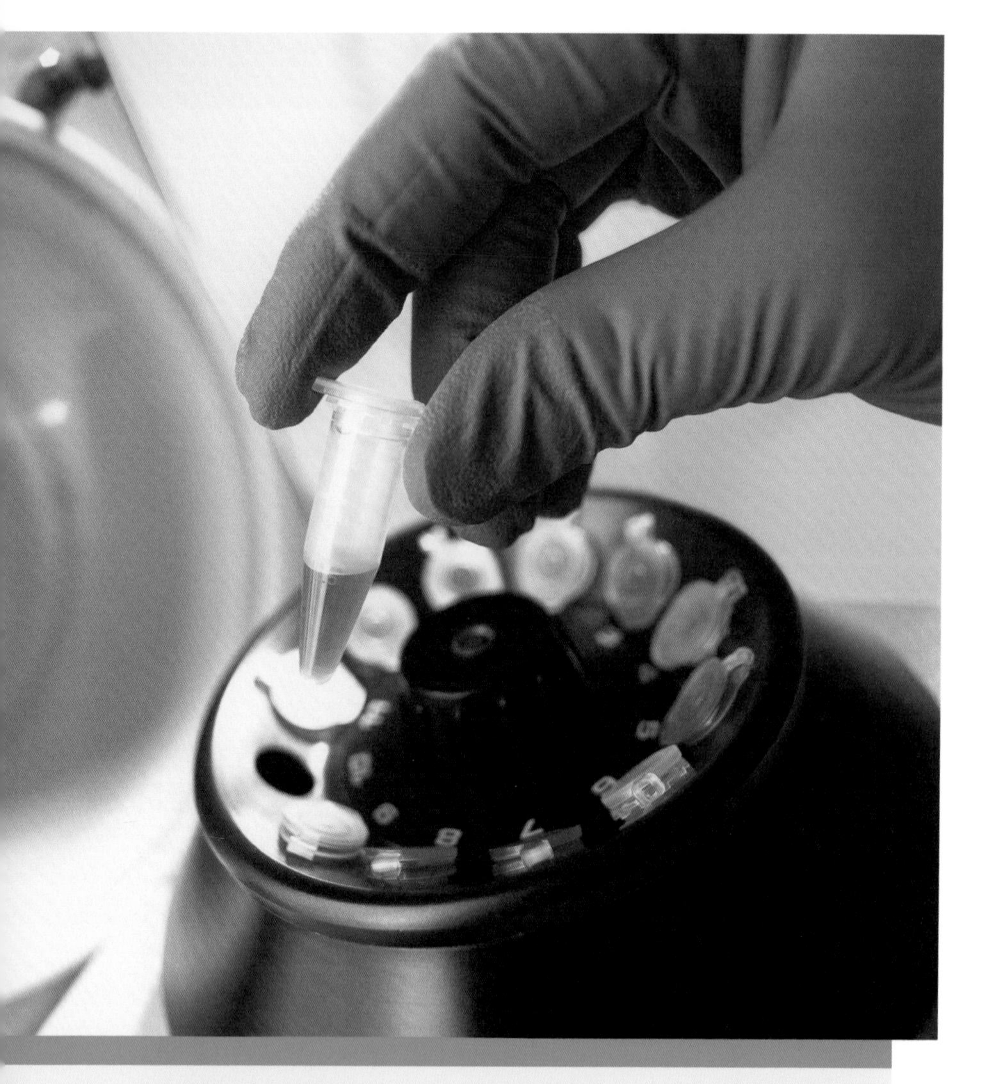

A laboratory worker inserts a sample containing DNA into a centrifuge; the rapid spinning of the centrifuge will separate the DNA from the liquid. *(Tek Image/Photo Researchers, Inc.)*

a variety of versions of each gene in the set. The Human Genome Project could sequence only a small number of samples, embodied in the DNA of individuals.

Due to privacy concerns, the project managers protected the identity of the individuals whose DNA was sequenced. Project workers collected samples from a large number of people from a variety of ethnic backgrounds. Samples of female donors came from blood, and samples of male donors came from semen, the male reproductive cells. Semen is the most efficient type of cell for DNA extraction because a high percentage of the material of these little cells is DNA, and the small size of the cell simplifies the process of purification (so that the final result will be pure DNA and not DNA mixed with other, unwanted material, such as proteins from the cell). Another advantage of using male cells is that they contain all the human chromosomes—females have two X chromosomes, so they lack the Y chromosome. But to maintain a balanced representation, DNA obtained from blood samples of females was also used. Only a small portion of the pool of donors provided the material for sequencing. Selection was random, and no one knows who donated the chosen samples.

Since everyone's DNA is different (except for identical twins), how was the sequence put together? A portion of one person's DNA might read GTACCA, but the same stretch of DNA from another person might be something entirely different. Sequencing the letters of DNA from multiple sources can be confusing.

But this problem was not a severe one, for scientists working on the Human Genome Project were already aware of the remarkable similarity of DNA among individuals. Genetic differences certainly exist—some people have one allele, others have another—but the differences are not great, even for people who are unrelated.

Some of the differences involve noncoding regions, especially in the number of repetitive sequences. Other variations arise from a difference in a single base. For example, a portion of DNA may be AAGATC in one person and AAGTTC in another—a T in the second sequence replaced the third A in the first sequence. This difference is known as a *single nucleotide polymorphism* (SNP), which people often pronounce "snip." A SNP in a coding region may result in a change in one of the protein's amino acids; for example, sickle-cell disease is caused by a change in a single base in the gene coding for a protein called hemoglobin—the codon GAG becomes GTG. GTG codes for valine and GAG codes for glutamic acid. This amino acid alteration, valine for glutamic acid, disrupts hemoglobin function and causes the disease.

Many people have heard the term *mutation* with regard to DNA changes or genetic differences among individuals. A mutation can be a large-scale change, such as a rearrangement of an entire chromosome, or a small-scale change, such as point mutation—a change in a single base. Scientists usually refer to DNA variations as mutations when they are rare. These changes may occur, for instance, when cells copy their DNA prior to dividing into two daughter cells, both of which require a full set of DNA. Mistakes can occur during the copying process. Cells have numerous mechanisms to correct these mistakes, but sometimes the mistake persists and, if so, will be transmitted to all of the cell's subsequent progeny. A mistake in the DNA of a skin cell, for example, would get passed on in the next division, and so on, resulting in a patch of skin that harbors the mutation. If a mistake occurs in a person's germ-line cell—a sperm cell or egg cell—and this cell goes on to form part of an embryo, the mutation will be carried by every cell in the newborn's body, for all cells derive from the fertilized egg cell. But such mutations will be rare because they will occur only in a few individuals. In terms of the entire population, a given mutation will be found in an exceedingly tiny fraction of people.

In contrast to mutations, SNPs are sites of common variability among the population. For a typical SNP, some percentage of the population has one base, say an A, and the rest has another, perhaps a G. Since there are four bases, a SNP may have up to four variations, but most have only two. There are millions of SNPs. Although they posed little problem for the sequencing project, SNPs are one of the most important aspects of the genome.

Even though variability did not greatly impede the sequencing process, three billion is a staggering number. No machine exists that can take this many bases at one time and sequence them. Even whole chromosomes are too large, so they had to be broken into smaller pieces. The process of breaking down molecules like DNA chromosomes does not yield a neat and orderly pile of pieces, but instead produces a jumbled heap. Researchers working on the Human Genome Project needed to find out where each piece belonged—on which chromosome, and where on the chromosome. To do this, researchers created maps consisting of small, widely spaced regions, called markers, whose locations were known. These markers allowed the identification of the pieces, since the markers of each piece told researchers where it belonged.

Several methods could be used for the sequencing procedure. The main method required breaking the pieces further down into smaller, overlapping fragments, which were small enough to be handled by the sequencer machines. A computer assembled the sequenced fragments into the proper order by using the overlaps to determine which fragments were neighbors. Over the course of the project, scientists and engineers discovered ways to improve the technology. Sequencing became faster and more automated. Four years were required to sequence the first billion bases, but the second billion took only four months.

In addition to the Human Genome Project, a private company, Celera, led by J. Craig Venter, initiated a separate project in 1998 to sequence the human genome. Celera's project employed a faster but more difficult approach to genome sequencing. Unlike the government project, which went piece by piece, Celera broke the whole genome into tiny fragments and used a sophisticated computer to assemble the huge number of sequences into the proper order. This process proved more rapid, although Celera did have access to the sequences already generated by the Human Genome Project to help it.

The Human Genome Project and Celera finished at roughly the same time. On June 26, 2000, then president William Clinton announced that a first draft of the entire sequence had been completed, calling it "the most important, most wondrous map ever produced by humankind." Researchers continued to refine the sequence and eliminate sequencing errors, and in April 2003 the job was essentially done. The Human Genome Project finished ahead of schedule, taking 13 years instead of the planned 15. In all, the cost of the Human Genome Project was $3 billion, although much of this money went for research associated with interpreting and studying the data instead of sequencing operations.

PHARMACOGENOMICS—PERSONALIZED MEDICATIONS

The completion of the Human Genome Project is just the beginning of the science of genomics. There is a huge amount of data to explore.

Now that the human genome has been sequenced, researchers who are studying a particular gene or DNA segment do not need to invest the time and effort to sequence it. But they still have to find it, which is easier than before the genome was sequenced, though it still requires some effort. Prior to the completion of the Human Genome Project, experts commonly estimated that the human genome contained about

100,000 genes, although only about 10–15 percent of these genes were known at the time. Having the whole sequence at hand, researchers can now sift through the genome and search for all the genes. Using start and stop signals for coding, as well as looking for sequences similar to known genes, scientists have identified about 20,000 human genes in all. This number is not exact because it is impossible to be certain that a DNA segment actually encodes protein or RNA simply from its sequence.

Identifying all the genes in the genome will take a little more time. Meanwhile, some researchers are focusing on the genetic differences among individuals. An important field of research that is particularly interested in SNPs and other variations is *pharmacogenomics*—the study of how genes affect a person's response to medicinal drugs. The study of drugs is known as pharmacology, so pharmacogenomics is a combination of pharmacology and genomics.

Physicians have long known that the response of individuals to a given drug can be highly variable. For some patients, a drug may work well, but for other patients, the same drug may be ineffective; similarly, a drug may invoke unintentional consequences called side effects in some patients but not in others. In some cases, the different responses may have to do with diet (food choices) or an interaction with other medications the patient may be taking, but in a large percentage of cases the difference in drug response is due to genetics.

For example, the most common cancer that strikes children is acute lymphoblastic leukemia. Doctors found drugs to cure the majority of children with this disease, but in some cases the drug not only failed to work, but also generated side effects that killed the patient. One of these drugs is 6-mercaptopurine. In the 1980s, researchers wanted to understand why drugs of this type, called thiopurine drugs, which were capable of making an astonishing cure in most patients, turned into a killer in others. Richard Weinshilboum, a physician and researcher at the Mayo Clinic in Rochester, Minnesota, and his colleagues discovered the answer to this riddle was in the genes. An enzyme called thiopurine methyltransferase, which prevents the drug from staying too long in the body, does not work properly in some people. As a result, the prescribed dose of 6-mercaptopurine was toxic to these patients. Weinshilboum developed a test that indicated the proper dose of 6-mercaptopurine based on the patient's gene that encodes this protein. This finding, and other similar research, led to an increased awareness of the role of genetics in pharmacology.

The extent of the problem of variable drug reactions has been documented in a 1998 paper, "Incidence of Adverse Drug Reactions in Hospitalized Patients," published in the *Journal of the American Medical Association.* Jason Lazarou, Bruce H. Pomeranz, and Paul N. Corey, of the University of Toronto in Canada, studied the frequency of harmful (adverse) reactions to medications that resulted in hospitalization or death. The researchers excluded incidents in which the medical staff made an error or the patient failed to observe instructions. Even with these exclusions, the number of harmful reactions was high. The researchers estimated that in 1994 there were two million serious reactions in hospitals in the United States, with more than 100,000 fatalities.

There is no reason to believe that estimates for other recent years will be much different. Although these numbers are based only on one research report, the data strongly suggest that harmful drug reactions affect large numbers of people every year. The cause in many of these cases is genetic.

Differences in responses to a drug can arise for a variety of reasons. Any drug, no matter what its intended effect, follows a basic course through the body. The drug is injected or absorbed into the body, and travels through tissue or the blood stream, where it interacts with one or more substances. The patient's body eventually eliminates the drug, often in the urine. At any point in its journey through the body, the drug may be altered in a chemical reaction with one or more of the body's molecules or enzymes. The alteration may be necessary for the drug's function, but it may also be unintended, causing a side effect. Side effects can also be caused when the drug interacts with healthy tissue, which was not the intended site of activity. (Drugs injected into the bloodstream, for example, travel throughout the body, not just to the diseased tissue or organ.) Since genes play an important role in controlling and regulating chemical reactions in the body, as in the case of thiopurine methyltransferase, a person's genetic makeup will affect how that person responds to medications.

To discover which genes are important for each drug, scientists need a great deal of data. In very few cases do physicians know which enzymes or other molecules react with a specific drug—most useful drugs are discovered either by accident or by trial and error, and the mechanism by which the drug works is unknown. Finding the right gene or genes means collecting two bits of information on a large group of patients: the patient's genetic makeup and how each patient responds to the drug. With this data, researchers can find correlations between gene variations and drug response; positive responders will have a particular set of genes (alleles) or gene regulators, and those who suffer adverse reactions to the drug will have another.

Because a large number of proteins may influence a drug's activity in the body, many genes could play a role in a patient's response. And because the human genome contains three billion bases, there is a lot of area to search. This is why pharmacogenomics researchers generally need a large number of patients. A small sample might fool researchers into thinking a minor genetic difference is important, when it really has little or no effect. If, for instance, researchers had data from only two patients, one of whom had an adverse reaction to the drug, then any genetic difference between those patients may be involved. Even though the average difference between two people is a little less than 1 percent, the list of possibilities would include hundreds of genes and thousands of polymorphisms. To home in on the right gene or genes, researchers need to compare data from a lot of patients. Only then will the list of possibilities get whittled down to one or a few genes or gene regulators that govern the response to that specific drug.

Although the Human Genome Project sequenced the entire genome, the data came from DNA of only a few individuals. This was not enough people to find all the SNPs and other genetic variations that may exist in the population. But in the course of their genetic research, many laboratories have sequenced genes or other DNA segments and have found variations among the test subjects. The National Center for Biotechnology Information (part of the NIH) has established a database, called dbSNP, containing these variations (which were contributed by the researcher or laboratory that discovered them). By 2009 the database included more than 6.5 million SNPs and continues to grow. Although this is a large number, it is much smaller than three billion—the number of bases—the SNP database allows researchers to focus their search on the common genetic differences among humans.

Even with SNP databases, pharmacogenomics research is complex. Patients often exhibit a wide range of responses to a drug—the response is usually not as simple as "get well" or "fail to get well," for there are side effects, time courses, and other factors to consider. In addition, researchers must have a patient's DNA data. This is not a trivial matter, for researchers must get the patient's permission to use this data. The patient's privacy must be respected, and the DNA has to be analyzed.

Genomic technology has yet to progress to the point where a person's entire set of chromosomes can be sequenced in a reasonable period of time. Until this or something similar can be done, researchers cannot make optimal use of databases, for they cannot identify all of a patient's SNPs and other variations. What generally happens is that researchers seek some sort of clue as to what kind of gene they are looking for, or where the gene might be located. For instance, there is a family of liver enzymes known as CYP that break down a large number of drugs, and are often good places to begin a search.

Other than the thiopurine methyltransferase success, and a few other cases, physicians and scientists have not made much progress thus far in pharmacogenomics. But this kind of research takes a great deal of time. As more genomic data accumulates, research in this field will become faster and more thorough, and progress in pharmacogenomics will accelerate.

GENES AND DISEASES

Genes not only influence a person's response to medications, they also influence, and sometimes cause, the diseases for which the medication is needed. Genetic diseases have been known for a long time—even the earliest physicians realized that some diseases "run in families," and therefore must be inherited.

Sickle-cell anemia is an example of a disease caused by a single faulty gene. The hemoglobin protein in people suffering from sickle-cell anemia does not function correctly because one of the amino acids is different. This change, or mutation, is inherited, but the trait—the disease—is recessive, which means that both copies of the gene must be changed for the trait to be expressed (in other words, for the disease to occur). When only one of the genes has this sickle-cell change, there is no sickle-cell anemia, but rather the person is more resistant to a disease called malaria, which is caused by a microorganism carried in mosquitoes. People who live in or are from tropical climates, such as parts of Africa and South America, are highly susceptible to malaria, so having one copy of the mutated gene is beneficial.

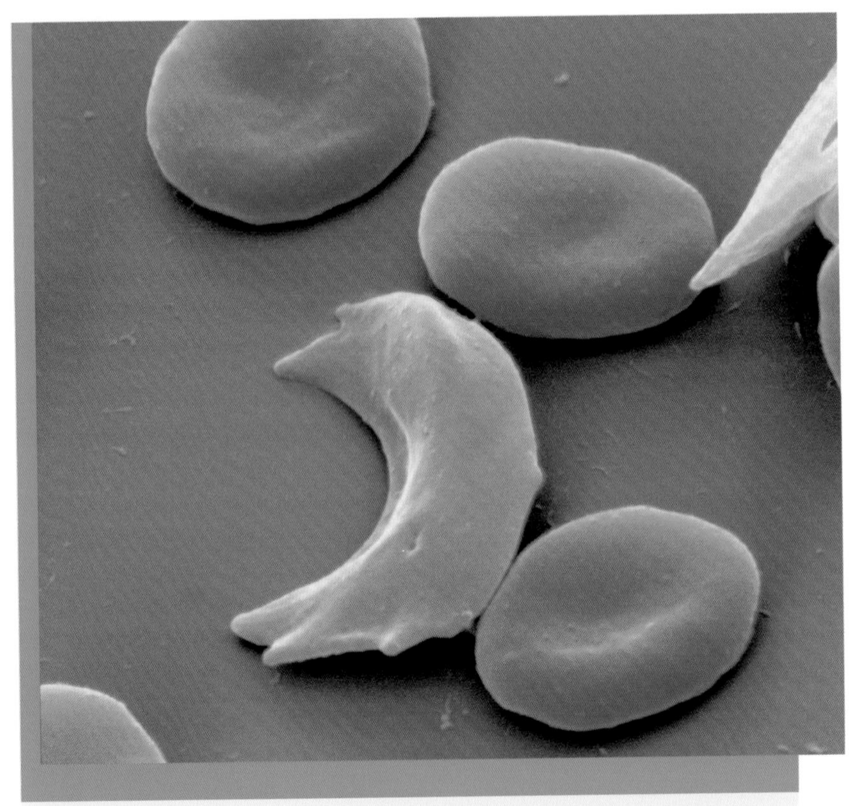

In sickle-cell anemia, red blood cells adopt a sickle shape, as compared to the oval shape of healthy cells. *(Eye of Science/Photo Researchers, Inc.)*

As is the case for most genetic diseases, there is no cure yet for sickle-cell anemia. But knowing which gene causes the disease is extremely helpful. Tests for the mutated gene can tell a person whether he or she carries the mutation; although one copy does not produce the disease, sickle-cell anemia will affect one in four children when both parents are carriers. This is because, on average, one child out of four will inherit both mutated copies, one from each parent. (One child of the four will likely inherit both normal copies, and the other two will have one—these two will be carriers. But the offspring of any one couple may not reflect this average, due to the chromosome and gene shuffling that occurs during human reproduction.) About 10 percent of African Americans are carriers, so these genetic tests are important.

Other tests exist for common genetic diseases such as cystic fibrosis. But the genetics for these diseases tend to be simple, involving only one gene. The relationship between genetics and other diseases is usually more complicated. For many diseases, genes may create a risk, or tendency, to get the disease, but do not necessarily cause the disease itself.

A prominent example is coronary heart disease, which is the leading cause of death in both men and women in the United States. Fifteen million Americans suffer from this disease, according to the American Heart Association. Coronary arteries are the small blood vessels that supply the heart with blood. As people age, fatty material as well as a substance called plaque can build up in the vessels, narrowing or even blocking the flow of blood to the heart and leading to a heart attack as heart cells die from a loss of nutrients. Physicians have identified several factors that contribute to coronary heart disease, including smoking, lack of exercise, and poor diet. But there are also other factors, some of which are beyond a person's control. Heart disease runs in families, and this means that there is a genetic component to the disease.

Finding the genes associated with coronary heart disease is not as easy as with cystic fibrosis and sickle-cell anemia. The relation between the disease and these genes is much more subtle; the genes may increase the deposit of fatty material, but this may be important only if the person engages in one or more high-risk behaviors such as smoking or having a sedentary lifestyle (in which the only exercise comes from lifting a fork to the mouth, or pushing the button of a television's remote control). So many factors are involved that physicians who study heart disease patients have difficulty identifying the associated genes. Experiments to single out specific factors by dividing people into rigorously controlled groups would not be ethical.

This is another one of the many areas where genomics can make an extremely important contribution. Having access to all of the genome, researchers have a better chance of discovering what they are searching for. But these studies generally require a long time, a lot of patients (and patience), and teams of researchers from different laboratories and universities to enroll and study the large number of patients required.

One important series of studies concerning heart disease has made genome-wide scans of patients in families with a history of the disease. Families are easier to search because investigators can look for markers, such as SNPs, that tend to get inherited along with the disease.

Conducting this research was the large team of Elizabeth R. Hauser and her colleagues at Duke University in North Carolina, along with researchers at GlaxoSmithKline in Research Triangle Park in North Carolina, Lausanne University Hospital in Switzerland, Vanderbilt University in Tennessee, University of Wales College of Medicine, Ludwigshafen Heart Center in Germany, University of Heidelberg in Germany, Baystate Medical Center in Massachusetts, Latter-Day Saints Hospital in Utah, Cardiovascular Associates in Virginia, and the Deborah Heart and Lung Center in New Jersey. The researchers identified a region on one of the chromosomes that was linked to the disease; this region tended to be inherited along with an increased susceptibility to heart disease. This result was published in a paper titled, "A Genome-wide Scan for Early-Onset Coronary Artery Disease in 438 Families," in a 2004 issue of the *American Journal of Human Genetics.*

In the next step, Liyong Wang, Elizabeth R. Hauser, Jeffrey M. Vance, and their colleagues at Duke University, along with researchers at Vanderbilt University, University of Wales College of Medicine, University of Sheffield, and the University of Miami, studied the SNPs in this region. They found a gene called kalirin that seems to play a role in the development of heart disease in these individuals. The researchers published their findings, "Peakwide Mapping on Chromosome 3q13 Identifies the Kalirin Gene as a Novel Candidate Gene for Coronary Artery Disease," in a 2007 issue of the *American Journal of Human Genetics.*

No one had previously suspected this gene. Having identified a prime suspect, researchers can now study the gene and its associated protein in an attempt to find out how it may contribute to the development of heart disease. Kalirin makes a protein that plays a role in the movement of cells within a type of muscle called smooth muscle, which forms part of the wall of blood vessels. Changes in this protein may cause cells to move too slowly or to collect in one spot, possibly creating a site where plaque can build up.

UNDERSTANDING THE HUMAN GENOME

The focus of most genome researchers has been on genes and genetic variation, which is appropriate given the vital role they play in develop-

Organisms pass along their genes to their offspring—and the offspring resemble their parents. *(Michael Sheehan/iStockphoto)*

ment, health, and disease. But only about 2 percent of the human genome contains coding sequences. What is the function of the other 98 percent? A lot of genomics is still labeled "Terra Incognita"—unknown territory.

Analysis of the human genome will greatly benefit from the sequencing of the genomes of other organisms. All life on Earth harbors similarities because of evolution; the genetic code, for example, is virtually identical among all organisms. By comparing the human genome with the genomes of other mammals, scientists can discover which sequences are the same and which are different or unique to a given species. Sequences that are the same are called conserved sequences. If a DNA sequence is not critical for an organism's survival, it is not maintained over long periods of time because natural mutations in this sequence do not adversely affect individuals. Conserved sequences, however, must be critical because they remain unchanged during evolution—individuals with mutations in these sequences do not survive.

The genomes of more than 1,000 organisms have already been sequenced. Many of these species are single-celled organisms, whose genomes are quite small compared to humans, but included in the list are the mouse, rat, dog, chicken, and chimpanzee. There is a lot of genetic similarity among all mammals, including mice and humans—most human genes have a counterpart in the mouse that is usually 70–90 percent similar in sequence. DNA of chimpanzees, the closest relatives to humans in terms of evolution, is about 97 percent identical to human DNA.

But the differences in genomes are also evident. Many organisms have little or no noncoding sequences, but mammals tend to have an abundance of such sequences, and humans have a great deal. Researchers realized soon after the discovery of the genetic code that much human DNA does not encode for genes. Japanese-American scientist Susumu Ohno was the first to term the seemingly wasted sequences as "junk DNA" in a 1972 paper, "So Much 'Junk' DNA in Our Genome," published in the book *Evolution of Genetic Systems.* Although the phrase *junk DNA* subsequently became popular, some researchers have stopped using it because at least some of the "junk" may have a definite use. Today people often refer to these sequences as noncoding DNA.

Some of the noncoding sequences are introns. Other noncoding sequences have a regulatory role, controlling which genes become expressed and when. Gene regulation often works by the binding or unbinding of molecules such as proteins to specific DNA sequences, most of which are located near but not within genes. Sometimes regulatory molecules prevent the transcription of genes by blocking the path of transcription enzymes. Other regulators enhance transcription by ensuring enzymes have access to the gene sequence. The function of most noncoding remains a mystery, yet genomic science is making progress on this issue, as addressed in the following sidebar.

In order to identify all of the functional segments in the human genome, the National Human Genome Research Institute launched a project in 2003 called ENCODE (which is short for Encyclopedia of DNA Elements). ENCODE provides funds to researchers interested in examining the human genome sequence for all genes and regulatory sequences, as well as sequences involved in chromosomal structure and replication. This effort aims to understand the genome as a whole by identifying and analyzing each of its parts.

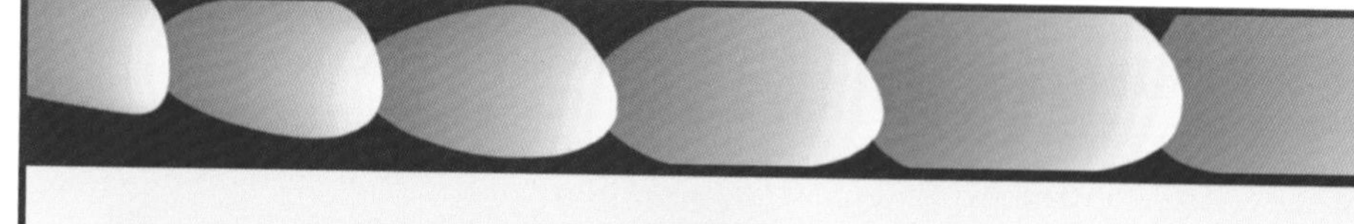

Noncoding "Junk" DNA

More than 50 percent of the human genome consists of repeating sequences. Some of this repetitive DNA consists of tandem repeats, which forensics experts often use to identify an individual's DNA, as described in an earlier sidebar. The number of bases in a tandem repeat can be up to about 10 bases, repeating itself dozens or even thousands of times in series. This DNA is sometimes referred to as satellite DNA, since it tends to separate from the rest of DNA during centrifugation (which involves spinning at high speed to separate a solution's components), forming a distinct region called a satellite band. A lot of satellite DNA is located at special chromosome sequences called centromeres and telomeres, which help maintain the structure of the chromosome. This satellite DNA may therefore play a structural role, which is less glamorous than the function of genes but still important.

Other repetitive DNA consists of repeating sequences scattered throughout the genome (unlike the tandem repeats, which are in series, repeating one after another). Most of these sequences seem to come from transposons—sequences that jump around. The exact function of this DNA in humans is not yet known, though they might be involved in gene regulation. Supporting this hypothesis is work published in *Proceedings of the National Academy of Sciences* in 2007 by Craig B. Lowe and David Haussler of the University of California, Santa Cruz, and Gill Bejerano at Stanford University in California. In the paper, titled "Thousands of Human Mobile Element Fragments Undergo Strong Purifying Selection near Developmental Genes," the researchers reported on a genome-wide scan for sequences that have been conserved over millions of years of evolution, as identified by their existence in a variety of mammalian species. What they found was a total of 10,402 DNA

(continues)

(continued)

segments, ranging in length from 50 to 489 bases and averaging 100 bases. Most of these segments came from regions of the genome where few genes reside ("junk DNA"), but they were close to regions that contain a lot of genes important for an animal's development. These conserved segments are likely crucial for survival, and their proximity to developmental genes suggests they exert some sort of regulatory influence, perhaps switching on and off certain genes as the embryo's tissue and organs grow and develop.

But other researchers have found evidence that at least some of the noncoding sequences in mammalian genomes are not essential. In a 2004 *Nature* paper, "Megabase Deletions of Gene Deserts Result in Viable Mice," Marcelo A. Nóbrega, Yiwen Zhu, Ingrid Plajzer-Frick, Veena Afzal, and Edward M. Rubin of the Joint Genome Institute and the Genomics Division of the Lawrence Berkeley National Laboratory in California studied mice with up to 1.5 million bases deleted from gene-free regions ("gene deserts"). The mice were born and developed with no problems, and were indistinguishable from normal animals in a variety of tests. This research suggests that at least some portions of noncoding sequences are, in fact, nonessential.

The initial phase of ENCODE spotlighted 30 million bases from 44 regions, constituting about 1 percent of the genome. In 2007, 35 research teams published a set of papers in *Nature* and *Genome Research* reporting their results. The findings were surprising. Scientists already knew of eight RNA-encoding sequences in these regions, but ENCODE researchers discovered many more. Even though coding sequences cover only a few percent of the genome, the ENCODE project teams found about 75 percent of the 30 million bases had been transcribed—an unexpected finding that has puzzled scientists. This transcribed RNA may have some as yet unknown function, or cells may normally dispose of it quickly. Another possibility that must always be considered is that the RNA transcripts are due to errors in the experimental procedure. But the researchers did not rely on a single procedure, and employed several labeling and sequencing methods to identify the RNA transcripts. More experiments are needed to study this newly discovered RNA.

ENCODE researchers also found an additional 200 regulatory sequences beyond what was already known in the regions under study. The proteins made from the genes in these regions were also more varied than expected.

The next phases of ENCODE will expand the investigation to include the whole human genome, as well as comparing the functional elements in human DNA with that of other species. Considering the unanticipated findings that have been reported thus far, the human genome probably has plenty more surprises awaiting scientists.

CONCLUSION

The genome is called the blueprint of an organism because the expression of genes in various cells and at various times is what makes the body and organs of a fly, a mouse, and a person. The completion of the Human Genome Project gave scientists access to the whole human blueprint, all in one package. But the sequence alone is not the whole story. The function of each segment must be understood, and individual variation, not documented in the Human Genome Project, is important in health, disease, and response to medications. Separate projects are ongoing to study these issues.

In addition to these projects, another important outgrowth of the Human Genome Project involves the study of proteins. Genes and gene regulatory sequences are crucial, but they are merely the means to an end—the molecules that carry out the instructions in cells and tissues are proteins and the various forms of RNA that are encoded by the genome. The study of the whole set of proteins of an organism is known as *proteomics*, a term that conveys a combination of protein science and genomics.

Unlike the genome, the number and amount of proteins in a living organism's cells and tissues varies over time. Responses to infection, injury, and stress require extra proteins of certain types to meet the challenges. The large number of proteins in cells and tissues, many of which interact with one another, presents a complex and variable picture for researchers to study. Although there are only about 20,000 genes—and only a subset of these are expressed in any one cell—many proteins in the human body undergo a variety of changes, such as the variations in RNA splicing described earlier, and chemical modifications such as the attachment of carbohydrates and other molecules. There may be as many as one or two million kinds of protein in a person's cells and tissues.

One recent study looked at the proteins associated with a substance known as high-density lipoprotein (HDL), a conglomeration of proteins and fats that helps remove dangerous buildup of cholesterol and other fatty molecules on the walls of blood vessels, which can result in serious health emergencies. (HDL binds to and moves cholesterol; the HDL-bound cholesterol is sometimes called the "good" cholesterol because it is carried away by HDL complexes.) Tomas Vaisar and his colleagues at the University of Washington, along with researchers at Wake Forest University in North Carolina and the Harvard Medical School in Massachusetts, analyzed the collection of proteins in HDL. In a 2007 paper published in the *Journal of Clinical Investigation,* "Shotgun Proteomics Implicates Protease Inhibition and Complement Activation in the Antiinflammatory Properties of HDL," the researchers discovered 48 proteins in the complex and noted a surprising number of HDL proteins that are also involved in the immune system's response to infection and the subsequent inflammation.

This study highlights the information that proteomics can yield as well as its complexity and challenges. Neither proteins nor genes act alone; genes function in concert and are regulated—expressed as needed—and their protein products can form large molecular assemblies that have a role in a wide variety of biological processes. The study and understanding of genomes and their products has accelerated rapidly since the sequencing of the human genome, but decades of research lie ahead.

CHRONOLOGY

1859 C.E. British scientists Charles Darwin (1809–82) and Alfred Wallace (1823–1913) propose the theory of evolution, in which some organisms outcompete others and survive to pass on their successful traits, thereby causing species to evolve.

1866 Gregor Mendel (1822–84), an Augustinian monk, publishes his experiments with pea plants and discusses how traits are inherited. His work is not well received.

1869 Swiss chemist Friedrich-Miescher (1844–95) becomes the first person to isolate and analyze DNA.

1900 German geneticist Carl Correns (1864–1933), Dutch botanist Hugo de Vries (1848–1935), and Austrian scientist Erich von Tschermak-Seysenegg (1871–1962) (whose grandfather had once taught Gregor Mendel) rediscover Mendel's work.

1902 American scientist Walter Sutton (1877–1916) and German zoologist Theodor Boveri (1862–1915) propose that chromosomes contain the units of inheritance.

1908 American geneticist Thomas H. Morgan (1866–1945) begins using fruit flies in his experiments in genetics. In a few years, Morgan and his colleagues show that units of inheritance reside on chromosomes.

1909 Danish botanist Wilhelm Johannsen (1857–1927) coins the term *gene.*

1928 British medical officer Frederick Griffiths (1879–1941) finds a component of bacteria that transforms certain traits.

1941 American scientists George Beadle (1903–89) and Edward Tatum (1909–75) propose "one gene–one enzyme" theory, in which each gene codes for a single enzyme. Although this theory is not quite right, it focuses attention on the issue of the genetic code.

1944 Canadian scientists Oswald Avery (1877–1955) and Colin MacLeod (1909–72) and American scientist Maclyn McCarty (1911–2005) show that genes are made of DNA.

1946 The Atomic Energy Commission (AEC), charged with studying reactions in nuclear physics, is born. Researchers begin to consider sequencing the entire human genome

1953 American biologist James Watson (1928–) and British scientist Francis Crick (1916–2004) propose that the structure of DNA is a double helix. They base their findings in part on experiments performed by British scientists Maurice Wilkins (1916–2004) and Rosalind Franklin (1920–58).

1960s Francis Crick, South African biologist Sydney Brenner (1927–), French biologist François Jacob (1920–), American scientist Marshall Nirenberg (1927–), and their colleagues decipher the genetic code.

1975 British biochemist Frederick Sanger (1918–) develops a method of sequencing DNA that is later employed in the Human Genome Project and elsewhere.

1976 Belgian biologist Walter Fiers and his colleagues are the first to sequence the complete genome of an organism, in this case a virus, MS2.

1986 DOE announces an initiative to sequence the human genome and begins funding some small projects aimed at developing the needed technologies.

1990 The Human Genome Project, sponsored by DOE and NIH, begins.

1998 Celera, a company headed by J. Craig Venter, launches its own project to sequence the human genome.

2000 President William Clinton, along with Francis Collins, the director of NIH's National Human Genome Research Institute, Ari Patrinos, chief of DOE's genome research, and J. Craig Venter, president of Celera, announce the completion of the first draft of the human genome sequence.

2003 Researchers finish the human genome sequence.

The National Human Genome Research Institute launches a project called ENCODE, which stands for Encyclopedia of DNA Elements, aimed at examining the human genome sequence for all genes and regulatory sequences

2007 ENCODE researchers publish their initial results, which indicate that more of the genome may be transcribed than scientists had earlier thought.

FURTHER RESOURCES

Print and Internet

Davies, Kevin. *Cracking the Genome: Inside the Race to Unlock Human DNA*. Baltimore: Johns Hopkins Press, 2002. The author provides in this book a detailed account of the science and technology as well as the politics and occasional showmanship involved in the sequencing of the human genome.

Department of Energy. "Genome Programs of the United States Department of Energy." Available online. URL: http://genomics.energy.gov/. Accessed April 1, 2009. The Department of Energy, along with the National Institutes of Health, has funded much of the research on the human genome. This Web page describes other department projects concerning microbial genomes, biofuels, pharmacogenomics, and many others.

DeSalle, Rob, and Michael Yudell. *Welcome to the Genome: A User's Guide to the Genetic Past, Present, and Future.* New York: Wiley, 2004. This well-illustrated book describes the human genome project and explores the medical and scientific advances that genome science may have in store.

Dolan DNA Learning Center, Cold Spring Harbor Laboratory. "DNA from the Beginning." Available online. URL: http://www.dnaftb.org/dnaftb/. Accessed April 1, 2009. With animations, galleries, and videos, this Web site offers a visual tour of DNA science.

———. "DNA Interactive." Available online. URL: http://www.dnai.org/. Accessed April 1, 2009. This informative and well-illustrated introduction to DNA, funded by the Howard Hughes Medical Institute, has sections about the history of DNA research, genomes, genetic engineering, and DNA applications.

Hauser, E. R., D. C. Crossman, C. B. Granger, J. L. Haines, C. J. Jones, V. Mooser, et al. "A Genomewide Scan for Early-Onset Coronary Artery Disease in 438 Families." *American Journal of Human Genetics* 75 (2004): 436–447. The researchers identified a region on a chromosome that was linked to heart disease.

Lazarou, Jason, Bruce H. Pomeranz, and Paul N. Corey. "Incidence of Adverse Drug Reactions in Hospitalized Patients." *Journal of the American Medical Association* 279 (1998): 1,200–1,205. The researchers report on the prevalence of harmful drug reactions.

Lowe, Craig B., Gill Bejerano, and David Haussler. "Thousands of Human Mobile Element Fragments Undergo Strong Purifying Selection near Developmental Genes." *Proceedings of the National Academy of Sciences* 104 (May 8, 2007): 8,005–8,010. The researchers report on a genome-wide scan for sequences that have been conserved over millions of years of evolution.

National Center for Biotechnology Information. "What Is a Genome?" Available online. URL: http://www.ncbi.nlm.nih.gov/About/primer/genetics_genome.html. Accessed April 1, 2009. This nicely written tutorial is an excellent introduction to the science of genomics.

National Human Genome Research Institute. "The ENCODE Project." Available online. URL: http://www.genome.gov/10005107. Accessed April 1, 2009. Sponsored by the National Human Genome Research Institute, the ENCODE Project aims to identify the functional elements in the human genome. This Web page provides news and information on this ambitious undertaking.

Nóbrega, Marcelo A., Yiwen Zhu, Ingrid Plajzer-Frick, Veena Afzal, and Edward M. Rubin. "Megabase Deletions of Gene Deserts Result in Viable Mice." *Nature* 431 (October 21, 2004): 988–993. The researchers studied mice with up to 1.5 million bases deleted from gene-free regions ("gene deserts"). The mice developed with no problems and were indistinguishable from normal animals in a variety of tests, suggesting that some portions of noncoding sequences are nonessential.

Oak Ridge National Laboratory. "Human Genome Project Information." Available online. URL: http://www.ornl.gov/sci/techresources/Human_Genome/home.shtml. Accessed April 1, 2009. This Web site provides a host of information on the Human Genome Project, including research, publications, ethical and legal issues, and educational resources.

Ohno, Susumu. "So Much 'Junk DNA' in Our Genome." In *Evolution of Genetic Systems,* H. H. Smith, ed. New York: Gordon and Breach, 1972, pp. 366–370. Susumu coined the term "Junk DNA" to describe regions of DNA that seem to lack functionality.

Richards, Julia E., and R. Scott Hawley. *The Human Genome: A User's Guide,* 2nd ed. Burlington, Mass.: Elsevier Academic Press, 2005. Although covering a lot of sophisticated material, this accessible book explains genes and genetics with a focus on health and disease issues.

Ridley, Matt. *Genome: The Autobiography of a Species in 23 Chapters.* New York: HarperCollins, 2000. Although somewhat dated (the year 2000 was a long time ago in terms of genome science!), this enjoyable book takes the reader on a tour through each of the 23 human chromosomes. No book can tell the whole story, so the author focuses on a single gene per chromosome and discusses its functions in nontechnical language.

Vaisar, Tomas, Subramaniam Pennathur, Pattie S. Green, Sina A. Gharib, Andrew N. Hoofnagle, Marian C. Cheung, et al. "Shotgun

Proteomics Implicates Protease Inhibition and Complement Activation in the Antiinflammatory Properties of HDL." *Journal of Clinical Investigation* 117 (2007): 746–756. The researchers analyzed the collection of proteins in HDL and discovered 48 proteins in the complex, noting a surprising number of HDL proteins that are also involved in the immune system's responses to infection and subsequent inflammation.

Wang, L., E. R. Hauser, H. S. Svati, M. A. Pericak-Vance, C. Haynes, D. Crosslin, et al. "Peakwide Mapping on Chromosome 3q13 Identifies the Kalirin Gene as a Novel Candidate Gene for Coronary Artery Disease." *American Journal of Human Genetics* 80 (2007): 650–663. The researchers found a gene called kalirin that seems to play a role in the development of heart disease in some individuals.

Watson, J. D., and F. H. C. Crick. "A Structure for Deoxyribose Nucleic Acid." *Nature* 171 (April 25, 1953): 737–738. Watson and Crick described the double helix structure of DNA in this paper.

Wellcome Trust. "The Human Genome." Available online. URL: http://genome.wellcome.ac.uk/. Accessed April 1, 2009. Founded by Sir Henry Wellcome and based in the United Kingdom, the Wellcome Trust funds research in a variety of medical and scientific fields. This Web site explores the human genome and contains dozens of features and articles about what scientists have learned and are presently researching.

Web Sites

Innocence Project home page. Available online. URL: http://www.innocenceproject.org/. Accessed March 30, 2009. This Web site offers news and information on the Innocence Project, which is devoted to the use of DNA technology to exonerate people who have been falsely convicted.

National Human Genome Research Institute. Available online. URL: http://www.genome.gov/. Accessed April 1, 2009. A branch of the National Institutes of Health, the National Human Genome Research Institute posts news and information regarding numerous research projects on its Web site.

3

Protein Structure and Function

Diabetes is a common illness, affecting about 20 million adults and children in the United States. What causes diabetes is an inability of the body's cells to take up enough sugar in the blood, resulting in hyperglycemia—excess sugar in the bloodstream. Unless properly controlled, diabetes can lead to heart disease, blindness, and other complications. To control the disease, patients generally receive injections of a substance known as insulin, which lowers blood sugar. Insulin is a hormone normally secreted by an organ called the pancreas, and signals cells that it is time to accumulate sugar—for instance, after a big meal. And like many molecules in the body that serve critical functions, insulin is a protein.

Prior to the 1920s, the disease could be controlled only by a strict diet. When researchers isolated insulin and showed that a deficiency of this hormone caused diabetes, treatment became possible by injecting insulin into the body. But there was a difficulty: What could be used as a source of insulin? The pancreas of certain animals also produces insulin, used for the same purpose as in humans, so early medical treatments involved harvesting insulin from pigs and cows. This insulin usually worked in humans but sometimes caused skin rashes and other problems because the animal hormone is not exactly the same as the human one. In the 1950s, British chemist Frederick Sanger (1918–   ) discovered the composition of insulin—the sequence of its amino acids—and a few years later chemists began to make it in the laboratory, but the quantity was limited. Then in the late 1970s, scientists at the City of Hope National Medical Center and the biotechnology company Genentech put the gene for human insulin into bacteria and tricked them into making a lot of human insulin protein. This and other similar techniques are the source for almost all insulin used in medical treatments today.

But *proteins* are large, complex molecules, and it is not always simple to get bacteria to produce a functional human protein. A large number of proteins besides insulin have been made by inserting the appropriate gene into bacteria, but sometimes, especially in the early years of this technique, the only reward researchers received for their trouble was an unusable mass of protein inside the bacterial cell. As scientists studied this problem, they began to realize that there was more to proteins than just their composition. Proteins serve many different functions in organisms, and these molecules have a structure—a shape—that is critical to the protein's function.

To learn how these vital molecules work, researchers need to study protein structure, but these molecules are too small to be viewed under ordinary microscopes. This chapter discusses the special tools and techniques that scientists use to visualize the structure of a protein. The discussion includes research aimed at finding out what happens when proteins do not adopt the correct shape, and the diseases that these faulty proteins can cause.

INTRODUCTION

Most people know about proteins in food. Protein is an essential component of a healthy diet, and is abundant in meat and also in a variety of vegetables. Consumed protein consists of protein molecules from plants or animals; a person's digestive system breaks down these proteins, and the body assembles the parts into its own proteins, as coded by its own set of genes. This is why insulin must be injected into the bloodstream rather than swallowed—the patient's digestive system would destroy it. The body often uses protein molecules to build up tissues, which is why bodybuilders like to include a lot of protein in their meals.

As discussed in chapter 2 of this book, most genes code for proteins, which are responsible for many of the functions of the body, such as hormones sending messages to cells. The importance of genes and proteins in health and disease prompted scientists to sequence the whole human genome—all human DNA—a monumental task completed in 2003.

Although this data provides a lot of information, there is still much to be learned about proteins. A new science has evolved, focusing on what has become known as the *proteome*—the whole set of proteins of an organism. (Researchers coined the term *proteome* to be similar to the term *genome*, the whole set of genes of an organism.) This topic includes structural genomics—the study of the structure of gene products such as proteins.

An emphasis on the whole set of proteins may be slightly ahead of its time, however. The human genome contains about 20,000 genes, which because of modifications and processing can produce many more different types of proteins—perhaps millions! Such a large number is daunting. While some scientists are willing to accept this challenge, other scientists continue to focus on certain individual proteins or small groups of proteins.

Many proteins are soluble in water. This means that they dissolve in aqueous (water) solutions, such as those found inside the body's cells (the intracellular solution) and outside (the extracellular solution). Similar to other soluble substances, water molecules separate these soluble protein molecules, which then float around in the water. It is often necessary for such molecules to move around in solution to perform their jobs; hormones such as insulin must travel in the bloodstream, for example, and soluble enzymes usually speed up chemical reactions by bumping into molecules and positioning them so that the reaction proceeds quickly.

Other proteins are insoluble, tending to form fibrous solid masses. For example, keratin is a fibrous protein in hair, and collagen is a protein found in connective tissue in the body. Neither of these proteins dissolve in water, which is a good thing, otherwise people would emerge from a shower bald and partially dissolved!

Solubility is an important factor—the proteins in bacteria "factories" mentioned above sometimes fail to form correctly, in which case the proteins that were supposed to be soluble instead aggregate and solidify. In order to function, proteins need to maintain a certain set of properties; a protein must adopt the correct form, and it must be stable, meaning that it does not change into some other form.

PROTEINS—CELLULAR MACHINES

The forms or structures of proteins must be remarkably varied because these versatile molecules perform a wide variety of jobs, including

Amino Acids

The name *amino acid* derives from two of the chemical groups that compose the molecule. Amino acids have a central carbon atom, to which four groups are attached: a hydrogen atom, an amino group (NH_2), a carboxyl group (COOH), and a side chain, which is different for each amino acid. Chemists have isolated more than 100 different amino acids, but only 20 are found in proteins. The figure shows the structure of these 20 amino acids. Note that proline has a slightly different structure, since its side chain is attached to the amino group.

In a neutral solution (not too acidic or too basic), which has a pH of 7, the hydrogen nucleus (a single proton) dissociates from the carboxyl group, leaving an electron and forming an acid, COO^-. This is where the acid in the term *amino acid* comes from. The amino group picks up a hydrogen nucleus and attains a positive electric charge.

The chemical nature of the side chain confers specific properties to the amino acid, and also to the proteins that are composed of them. Some side chains are electrically charged (in neutral solutions), some are *hydrophilic,* meaning they are attracted to water molecules, and some are *hydrophobic,* meaning they are repelled by water. The interaction with water comes about because water is a polar molecule—although an intact water molecule is electrically neutral, the oxygen nucleus has a stronger pull on the shared electrons of the covalent bond between oxygen and hydrogen, resulting in an unequal distribution of electrical charge. This creates an electric field, and polar molecules are attracted to each other because of it. (The field is also the basis of hydrogen bonds.) Polar molecules are therefore hydrophilic, as are charged molecules. A hydrophobic molecule is nonpolar, having little or no electric field, and nonpolar molecules congregate together when faced with polar molecules.

Glycine (Gly)	Alanine (Ala)	Valine (Val)	Leucine (Leu)	Isoleucine (Ile)
Serine (Ser)	Threonine (Thr)	Phenylalanine (Phe)	Tyrosine (Tyr)	Tryptophan (Trp)
Aspartate (Asp)	Glutamate (Glu)	Asparagine (Asn)	Glutamine (Gln)	Cysteine (Cys)
Methionine (Met)	Lysine (Lys)	Arginine (Arg)	Histidine (His)	Proline (Pro)

The structure of each of the 20 amino acids follows the same pattern, consisting of a central carbon molecule with four attachments: a hydrogen atom shown on the right, carboxylic acid (COO^-) on top, amino group on the left, and the specific side chain on the bottom.

hormones that act as messengers, enzymes that catalyze specific reactions, and insoluble proteins such as keratin or collagen that provide covering and support. Other proteins, such as kinesin and dynein, work as "motors" to transport molecules throughout the cell. Moving along fibers in a cell, kinesin and dynein play a vital role in escorting certain molecules where they need to go, which is much more effective than relying on the random movements that molecules normally make. Kinesin travels about 10 inches (25 cm) a day—not exactly fast, but it does not usually have to travel far.

In the first decade of the 20th century, German chemist Emil Fischer (1852–1919) showed that proteins are composed of smaller molecules called amino acids. A protein is a chain of amino acids, with each amino acid connected to the next member of the chain by a strong covalent bond known as a peptide bond. Proteins are therefore polymers, composed of repeating units (amino acids). The sidebar on pages 80–81 describes the 20 different amino acids found in proteins.

Short chains consisting of only a few amino acids are called peptides. Most proteins contain 100–300 amino acids. The largest protein in the human body is titin, a gigantic protein in muscle that has 27,000 amino acids. British chemist Frederick Sanger (1918–) pioneered the technique to determine the amino acid sequence of a protein. In the early 1950s, Sanger developed an elaborate set of chemical reactions that allowed him to discover the order of amino acids in insulin. Most researchers use a different technique today to find a protein's sequence, as discussed below.

The structure of a protein is critical to its function, but is the sequence related to the structure? Researchers in the 1950s were unsure. But a decade later, American chemist Christian Anfinsen showed that proteins rely on the sequence to adopt their shape; in other words, the order of amino acids is critical.

What Anfinsen did was to conduct a series of experiments in the 1960s to see if a protein readily adopts its proper shape. In an abnormal solution, such as an extremely acidic (low pH) or basic (high pH) solution, proteins lose their functional properties because they lose their shape—they "unfold." But when Anfinsen introduced these nonfunctional proteins into the proper solution, the proteins regained their proper form.

The shape or form is called the *conformation* of the protein, and Anfinsen showed that proteins adopt their conformation spontane-

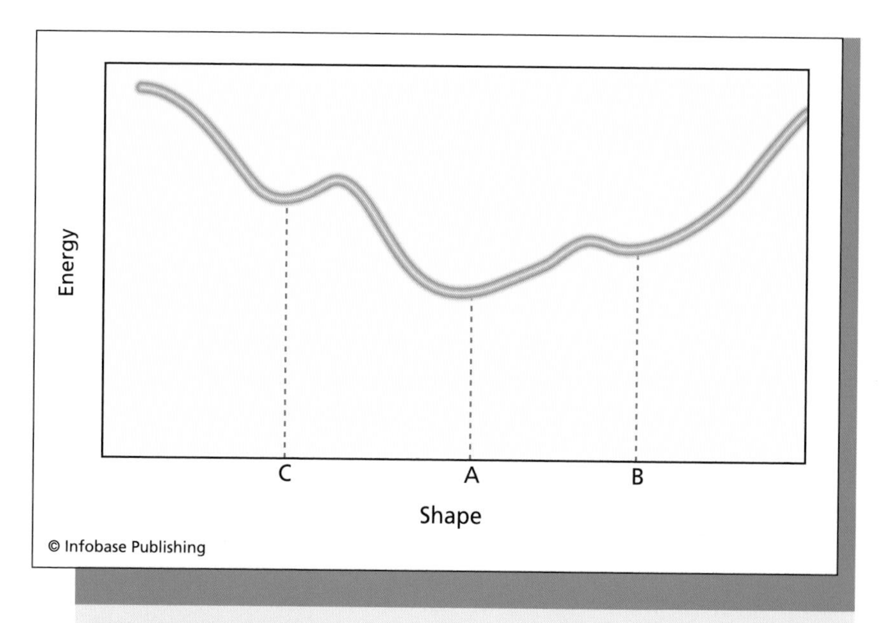

For each possible shape of the protein, specified by the horizontal axis of the graph, the corresponding energy is plotted on the vertical axis. The minimum is at A, the shape into which the protein will generally fold. But dips in the graph, such as at points B and C, are different shapes at which the protein could get "stuck," and fail to reach the overall minimum.

ously. A natural or spontaneous event occurs when an object or system proceeds to its lowest state in terms of energy. For instance, a ball rolls downhill, coming to a stop at the lowest point, the state that minimizes its gravitational potential energy—the energy it possesses because of its height. A protein folding into shape is a similar process, although its energy comes from chemical or electrical interactions rather than from gravity. Scientists often draw a "landscape" to model the process of protein folding, in analogy to a falling ball; the protein "falls" to its shape because it "rolls" to the lowest level, as shown in the figure.

Since proteins fold naturally, the information to do so correctly must be contained within the protein. This information is related to the sequence of amino acids. Peptide bonds join the amino acids in the chain, but these bonds do not create the shape—even when a protein loses its structural form, it still consists of a sequence of bonded amino acids. Instead, the shape comes from other bonds and interactions of the amino acids that are governed by the order of the amino acids in the chain.

PROTEIN STRUCTURE AND CONFORMATION

Most proteins fold into their correct shape automatically if conditions allow it; this is what Anfinsen discovered in his experiments. Proteins lose their conformation when their environment—the solution in which they are dissolved—pulls them out of shape. Such unfolding occurs when the concentration of salts in the solution is too high or too low, when the solution has the wrong pH, or even when proteins are exposed to high temperatures. Heat causes unfolding because heat is energy and is associated with greater movement of atoms and molecules. When the temperature is high enough, the atoms of a protein jiggle around so much that they pry apart the shape. The body's physiology and chemistry maintain a narrow range of temperature (the human body is usually within a degree of 98.2°F, or 36.8°C), concentrations, and pH (the normal pH for blood is about 7.4).

Although proteins do not require a mold or a set of instructions to fold, many proteins have help. Molecules called chaperones assist the folding process in various ways, such as by placing the protein into the best position for it to fold, or by preventing unfolded proteins from sticking together. Unfolded proteins, also known as *denatured* proteins, sometimes form a mass that precipitates out of the solution, settling into a solid clump. This is what happened in many early attempts to produce human proteins using bacteria, and it is also what occurs when egg white forms in a boiled egg. Chaperones are proteins (and some chaperones use other chaperones to help them fold!) and are also known as heat shock proteins because they are called upon to help when heat threatens to pull proteins apart.

Despite the help of chaperones, proteins do not need to be "told" how to fold—all the information needed to assume the proper shape is contained in the molecule itself. For a protein, this means its sequence of amino acids, which is governed by the gene that encodes them.

Chapter 2 of this book discussed how genes code for proteins. The DNA sequence of the gene is a code that enzymes read and use to build the protein, positioning the amino acids for peptide bonding. The bases of DNA are adenine (A), thymine (T), guanine (G), and cytosine (C), and three consecutive bases form a codon that codes for a single amino acid. For instance, GGC codes for the amino acid glycine. The information in the DNA molecule is first copied to a specific type of RNA molecule known as mRNA (messenger RNA). Then the mRNA binds to a ribosome—a huge complex of proteins and RNA. Ribosomes read a codon, attach the correct amino acid to the growing chain, and move to the next codon. (See the figure on page 46.) The result is a string of amino acids in their proper order as dictated by the gene sequence.

The sequence of a protein's amino acids can therefore be determined by the DNA sequence of the gene. This is how researchers today usually determine a protein's amino acid sequence, instead of using an elaborate set of chemical reactions to determine the order of the amino acids in the protein itself. Thanks to the Human Genome Project, biologists often find the gene before they find the protein.

Amino acids differ only in their side chains. Scientists quickly realized that the side chains, among other groups in the amino acids, must interact with each other and with molecules in the solution in order to form the bonds that stabilize the protein in the proper conformation. These bonds are not usually as strong as covalent bonds, such as peptide bonds, or ionic bonds, such as the bond between sodium and chloride that forms sodium chloride (table salt). A protein's shape is not rigidly welded with strong chemical bonds, but is instead made up of a large number of weaker interactions among its components, giving proteins flexibility and adaptability.

Attractions and repulsions of the electric charges of amino acid side chains also play a role in stabilizing the shape, along with much weaker interactions that are sometimes called van der Waals forces. Named after Dutch physicist Johannes Diderik van der Waals (1837–1923), these forces are relatively weak attractions between atoms and molecules due to uneven or fluctuating electric charge distributions. Hydrogen bonds, which play an important role in stabilizing DNA molecules, form because of an electrical attraction between a hydrogen atom and atoms such as oxygen and nitrogen. Heat or highly acidic or basic solutions pry these weakly attracted atoms apart, denaturing the protein.

In addition to these weak interactions, sometimes a strong covalent bond called a disulfide bond forms between cysteine amino acids. This bond forms between the two sulfur atoms in the side chains of two cysteine amino acids, helping to lock the protein's shape in place.

One of the most important stabilizing interactions is the hydrophobic nature of some of the amino acids, as discussed in the sidebar above. Many proteins have a hydrophobic region that curls into a ball in the central core of the protein, much like a drop of oil or fat will stay together in a bowl of water. Unfolding would expose the core to water, so the protein tends to stay folded up.

What sort of shape do these interactions impose on proteins? A hydrophobic core might be expected to give a protein a globular shape, but otherwise it is not obvious what sort of geometry a folded protein could have. The question is not an easy one to answer because, even though proteins are large molecules, they are not large enough to be seen by the unaided eye or the best optical microscopes. Visible light is much too "big" for such studies—the wavelength of light that people can see exceeds the size of a protein, which means a single protein molecule cannot reflect this light. Instead, scientists use a type of electromagnetic radiation with a much smaller wavelength and higher frequency than visible light—X-rays.

X-rays have wavelengths in the range of $0.4 \times 10^{-6} - 0.4 \times 10^{-9}$ inches ($10^{-6} - 10^{-9}$ cm). These fast oscillations give X-rays enough energy to pass right through the body, or, if striking an atom or molecule, to knock electrons out of their orbits and break bonds, which is why exposure to a large dose of X-rays is dangerous to living organisms. Physicians use a small amount of X-ray radiation to image bones inside the body—the heavy atoms such as calcium in a bone tend to absorb or block X-rays, unlike the soft tissues of the body that do not impede much of the radiation—so bones create a "shadow" on an X-ray film.

Researchers use X-rays to study the structure of molecules because this radiation's wavelength is the right size to "illuminate" the tiny distances between atoms. As described in the following sidebar, using X-rays to determine structure requires a lot of molecules situated in a precise arrangement, known as a *crystal.* Protein crystals used for these studies tend to be strongly hydrated—they contain many water molecules—so the protein's structure in the crystal does not differ much from the one it normally adopts in solution.

In 1958, British biochemist Sir John Cowdery Kendrew (1917–97) and his colleagues were the first to use X-ray analysis to determine the structure of a protein. The protein was myoglobin, found in muscle (the

X-ray Crystallography

Scientists and engineers use X-ray crystallography to study many different molecules. The idea to use X-rays to study molecular structures arose shortly after German physicist Wilhelm Röntgen (1845–1923) discovered X-rays in 1895. German physicist Max von Laue (1879–1960) suggested the procedure in 1912, and at about the same time, British scientists Sir William Henry Bragg (1862–1942) and his son, Sir William Lawrence Bragg (1890–1971), deduced the mathematical framework for the procedure. British chemist Dorothy Crowfoot Hodgkin (1910–94) pioneered the technique for the study of large biological molecules, and in 1956 succeeded in discovering the structure of vitamin B_{12}. A few years earlier, British scientist Rosalind Franklin used X-rays to probe DNA, leading to the discovery of the double helix by American biologist James Watson and British scientist Francis Crick.

Both the eye and a camera create images by using a lens to focus light, but X-rays are so energetic that they are difficult to focus. Instead of a lens, X-ray crystallography makes use of diffraction, a property of electromagnetic radiation in which the path of the radiation becomes distorted or spread out when it passes by the edge of a surface or through a small hole. For example, the tiny tracks of a CD or DVD diffract light and result in a spectrum of colors. In X-ray crystallography, X-rays irradiate the sample under study, and the radiation is diffracted and scattered, forming a pattern on a detector. The pattern is a set of spots that do not look anything like the structure of the molecule. The figure on page 88 illustrates an example of an X-ray diffraction pattern. In order to interpret the pattern, scientists apply advanced mathematical techniques

(continues)

(continued)

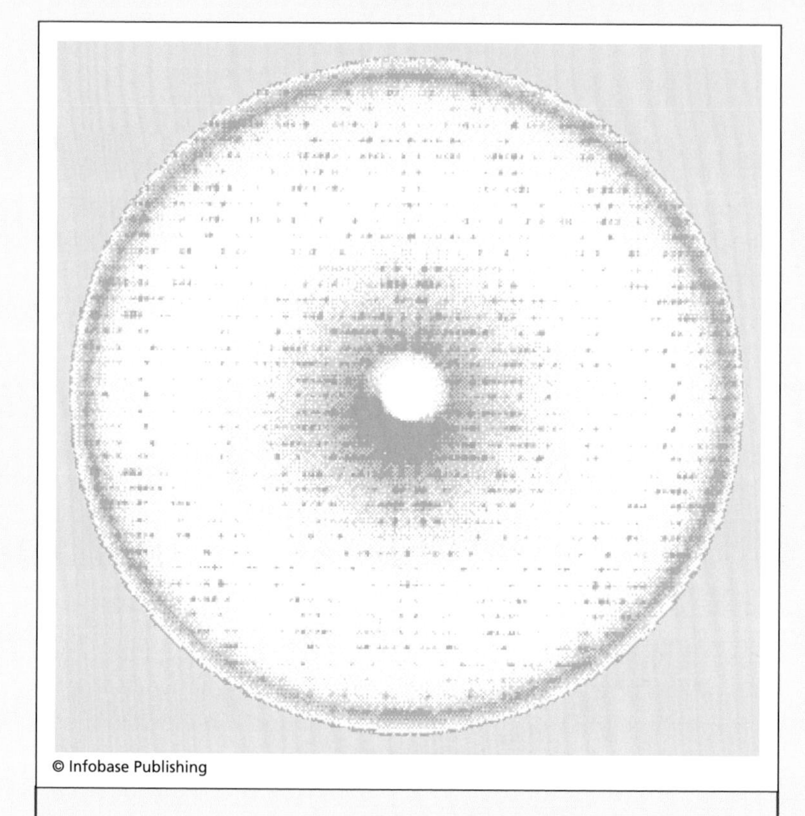

The protein crystal diffracts X-rays, which then strike the detector or a plate of X-ray film and make a pattern of spots. Scientists must apply mathematical techniques such as the Fourier transform to determine the shape of the protein from this pattern.

tissue used by Kendrew and his colleagues came from sperm whales, which have a lot of muscle!). Since 1958, many protein structures have been determined and made available, kept in large public databases. The Protein Data Bank, begun in 1971 with a list of seven structures, holds such as the Fourier transform to reconstruct what the structure must be, in order to have produced the observed X-ray diffraction pattern. Many computer programs have been written to compute the Fourier transform (named for French mathematician Jean Baptiste Joseph Fourier [1768–1830]), which is useful in almost every area of science and engineering.

The cause of the diffraction in X-ray crystallography is electrons in the atoms that compose the molecule. The Fourier transform converts the diffraction pattern into a map of electron density—congregations of electrons. Electrons surround the atomic nucleus, so researchers can determine from these maps the location of the atoms, with the exception of hydrogen—having only one electron, hydrogen is difficult to see with this technique. Each pattern produces a flat, two-dimensional map, but by sending X-rays through a variety of angles, researchers can visualize the molecule's three-dimensional structure.

Using this procedure on a single molecule would not produce a detectable signal, since there would not be enough scattered radiation. A lot of molecules are required, but unless all of them have the same orientation, the X-rays will be scattered randomly and the pattern will not be decipherable. Crystals are composed of a geometric arrangement of repeating structures, and this repetition is what gives researchers the uniform signal they need. Researchers must grow the crystals under special conditions, and any impurity (foreign molecule) or imperfection in the crystal degrades the data.

more than 56,800 structures from a variety of organisms as of April 2009. About 85 percent of these structures come from X-ray studies; the rest involve other techniques, such as *nuclear magnetic resonance* (NMR) spectroscopy, which will be described later.

This nuclear magnetic resonance spectrometer is located at the Pacific Northwest National Laboratory. *(EMSL at Pacific Northwest National Laboratory/U.S. Department of Energy Office of Science)*

Although many proteins are roughly globular, the amino acids do not just wind around the center like a ball of string. Stretches of amino acids often adopt repeating structures, the most common of which

(opposite page) The primary structure of a protein is its amino acid sequence. A protein's chain of amino acid folds into a secondary structure, such as an alpha (α) helix or beta (β) sheet, which in turn folds into the overall shape, known as the tertiary structure. Sometimes certain folded protein chains bind together, forming a bigger protein that has a quaternary structure—the shape of the conglomeration.

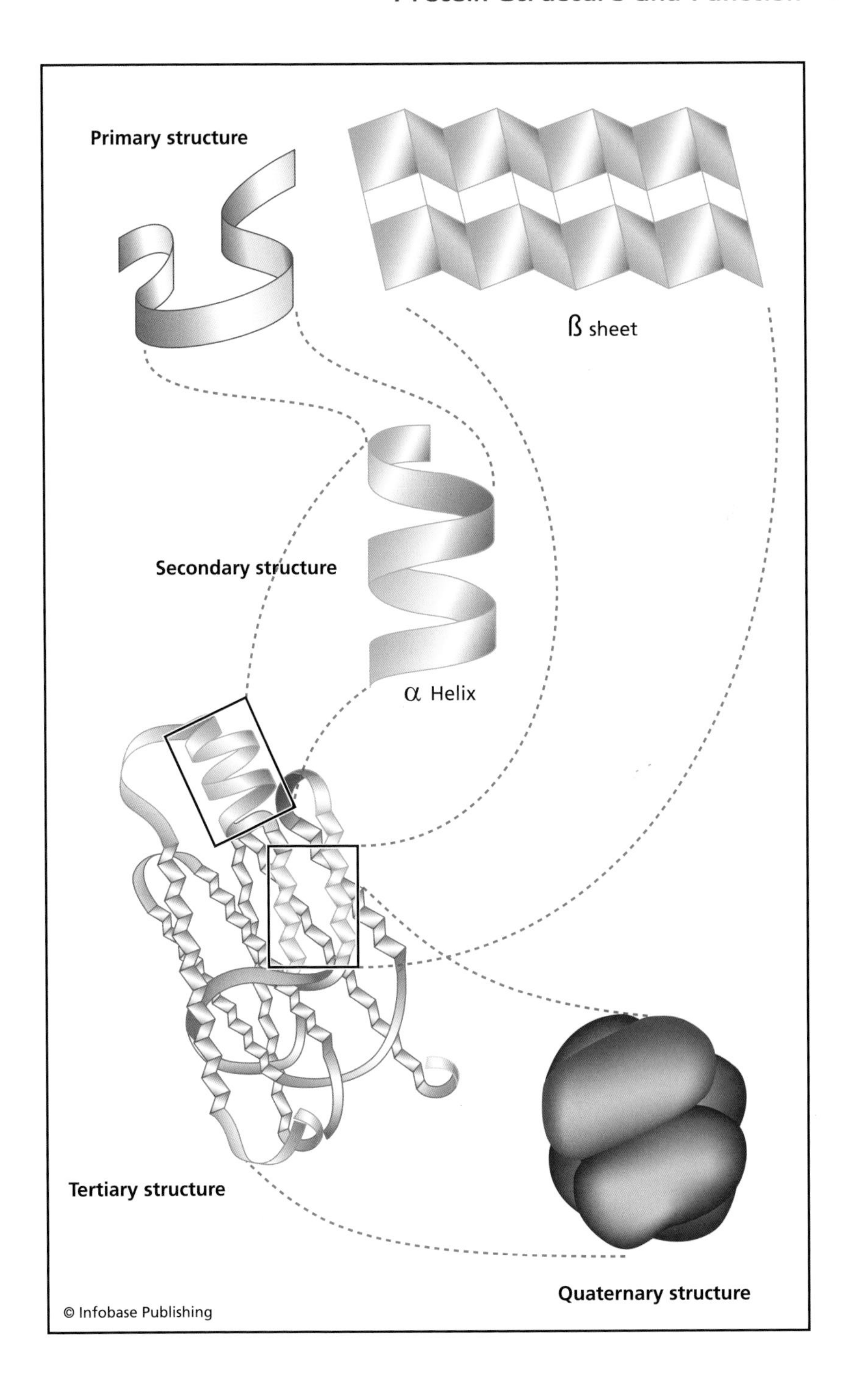
Primary structure
ß sheet
Secondary structure
α Helix
Tertiary structure
Quaternary structure
© Infobase Publishing

is known as an alpha helix. In this structure, stabilized by hydrogen bonds, the amino acid chain curves around a central axis, similar to a DNA helix but not, of course, double-stranded. The figure on page 91 shows an example. Each turn consists of three amino acids plus a little more than half of a fourth one. (The term *alpha* is the first Greek letter, and was used by early researchers to refer to a specific structural form, which has since been identified as a helix.) Another common structure is known as the beta sheet, as shown in the figure. This structure is an assembly of pleated sheets that is also stabilized by hydrogen bonds.

Most insoluble proteins are fibers that consist mainly of alpha helices or beta sheets. Soluble proteins also have these repeating structures, but in most soluble proteins the helices and sheets fold into a globular shape, keeping a hydrophobic core away from water molecules. Scientists refer to the alpha helices and beta sheets as a protein's secondary structure, and the globular folding is the tertiary (third) structure. In some proteins a number of separate chains bind (weakly) together, forming a quaternary structure. All of these structures are governed by the position and order of amino acids, which is the primary structure.

Knowing the structure of a protein means that researchers can often gain a better understanding of how the protein functions. For example, many enzymes speed up chemical reactions by binding the reactants—the participating molecules—bringing them close enough for the reaction to occur. Enzymes typically require a special shape to do this, and must have sites on the protein where the reactants bind. By examining the structure, researchers often get clues as to how the protein works as well as some ideas of how to improve or adjust the design.

Although a large number of protein structures have been determined, there remains a great deal of work to do. Some proteins are embedded in cellular structures such as the membrane—the cell's lipid (fatty) outer shell—or are bound to other molecules, which complicates their structures and makes them difficult to determine. Proteins that span the membrane are exceptionally difficult because they rarely form crystals, and few such structures have been determined. This is one of the reasons why protein structures is still an exciting field of study.

In 2006, Clare A. Peters-Libeu, Yvonne Newhouse, Danny M. Hatters, and Karl H. Weisgraber of the Gladstone Institute in San Francisco, California, and the University of California, San Francisco, found the structure of an important protein called apolipoprotein E (abbreviated

Isolating a protein under study, as these researchers are doing, is an important step in determining its properties. *(Argonne National Laboratory Media Center)*

apoE) while it was bound to a lipoprotein (a molecule containing fats and protein). This protein transports molecules such as certain vitamins and cholesterol, and plays an important role in cardiovascular health. The gene for this protein has also been found to be involved in Alzheimer's disease, which attacks the brain and produces irreversible memory and cognitive deficits. Researchers do not know what role apoE plays in Alzheimer's disease, but a certain allele of the gene greatly increases a person's risk of contracting the illness. Although the structure of apoE had been previously determined, the researchers' paper, "Model of Biologically Active Apolipoprotein E Bound to Dipalmitoylphosphatidylcholine" published in *The Journal of Biological Chemistry,* was the first to use X-ray analysis to construct a model of the protein in the performance of one of its duties.

OUT OF SHAPE: PROTEIN MISFOLDING

Proteins cannot function correctly unless they are properly folded. In addition to affecting its solubility, as discovered by early researchers who used bacteria as protein factories, improper folding can cause problems in a lot of different ways. As Vladimir N. Uversky of the Indiana University School of Medicine and his colleagues wrote in a 2006 issue of *Journal of Proteome Research,* "[I]t is recognized now that many devastating disorders, including neurodegenerative diseases, amyloidoses, cataracts, arthritis, and type 2 diabetes, belong to the family of so-called protein misfolding or conformational diseases." One of the most prominent human disorders involving protein misfolding is Alzheimer's disease.

Alzheimer's disease takes its name from the German physician Alois Alzheimer (1864–1915), who first described it in 1907. This disease is a brain disorder, and in its most common form, the disease tends to strike people over the age of 65, initially resulting in bouts of forgetfulness and confusion, and then progressing to a nearly total loss of mental faculties. More than five million Americans are afflicted. Physicians have not yet found a cure.

There is a genetic component to this disease, as in most diseases—a certain version of the gene coding for apoE, mentioned above, elevates a person's risk. Environmental factors such as pollutants and chemicals may also play a role in contracting the disease, and the risk of Alzheimer's also seems to increase for people who have inactive minds (such as people who spend excessive amounts of time watching television). But Alzheimer's is primarily associated with an accumulation in the brain of plaques and tangles, both of which are dense and abnormal aggregations of improperly folded proteins.

Plaques form in the spaces between brain cells, and contain amyloid beta, a short peptide whose normal function is not yet clear. Usually soluble, at high concentration these peptides change shape and produce fibrous structures. Tangles, found within brain cells, are made of a protein called tau, which normally forms part of the internal structure of a cell but in Alzheimer's disease aggregates into unwieldy masses. The brains of most older individuals have at least a few plaques and tangles, but in Alzheimer's patients, plaques and tangles proliferate. As the disease progresses, brain cells eventually die, causing massive tissue loss in the brain.

No one is sure why plaques and tangles form, nor are all researchers convinced that these misfolded proteins are the primary cause of the disease. It is possible that some factor, as yet unknown, is responsible for the death of brain cells and also produces, as a side effect or as part of the process, a large number of plaques and tangles. Research into Alzheimer's disease is hampered because only humans contract the disease; since laboratory animals do not normally get Alzheimer's, experimental opportunities for biologists are limited. Scientists and physicians have focused on the proteins involved, and efforts to understand their structures and functions are ongoing in many laboratories.

A critical aspect of any misfolded protein is determining how it got that way. The folding process might be expected to be simple if the energy "landscape" is steep and smooth, in which case folding should proceed without delays. Like a ball rolling down a hill, the protein "falls" into its proper shape. If the landscape contains ledges or little valleys, the protein may get stuck and fail to get all the way to the end—in other words, it will not be properly folded. (See the figure on page 83.)

Most small proteins fold within a fraction of a second, often around a thousandth of a second. Larger proteins, and proteins that are assembled or inserted into specific locations, can require a much longer time. But even a small protein with, for instance, 20 amino acids, has a fantastically large number of possible shapes, if the protein's components can move in any way or adopt any orientation. American biologist Cyrus Levinthal pointed out in 1968 that proteins do not have time to go through all possible shapes in order to find the right one, so the folding process must be direct. This leads scientists to believe that the energy landscape is in most cases steep and smooth, directing the proteins to the proper shape with few or no intermediates. In other words, an unfolded protein snaps into shape as a whole; there is no step A, then step B, and so on but simply one giant leap to the finish line. Some proteins seem to behave exactly in this way, for solutions of many proteins are

This animal, afflicted with bovine spongiform encephalopathy (mad cow disease), has difficulty remaining upright and may suffer from a variety of other symptoms, such as behavioral aggression and significant weight loss. *(U.S. Dept. of Agriculture—Animal and Plant Health Inspection Service)*

what scientists call a two-state mixture. The proteins in these solutions have only one of two possible shapes: fully folded, which is stabilized by bonds and interactions, and unfolded.

But there are exceptions. Some proteins have stable intermediates—a partially folded state that persists, stabilized by a smaller but still considerable number of bonds. A protein may get stuck in this state and become unable to progress to the fully folded state, in which case it does not function appropriately. In some cases this can be a disaster, as with a molecule known as a *prion*.

Diseases associated with prions are encephalopathies, which attack the brain of humans and animals (the term *encephalopathy* comes from the Greek *kephalē,* head, and *pathos,* suffering). The list includes bovine spongiform encephalopathy (BSE, also known as mad cow disease) and, in humans, Creutzfeldt-Jakob disease. Symptoms of Creutzfeldt-Jakob disease start with memory loss, confusion, and hallucinations, with a progressive loss of mental function as brain cells die. These fatal diseases are rare, but there is much concern about humans eating meat tainted with BSE. Although there is still some debate, the vast majority of scientists believe that prions cause these diseases.

The term *prion* stands for proteinaceous (protein-containing) infectious particle. A prion is a specific protein that is not folded correctly, and has the extremely unusual property that it can bind to correctly folded proteins and induce them to adopt the malformed shape, thereby propagating the disease-causing conformation. The normal protein is known as PrP^{C}, a soluble protein containing 231 amino acids found in membranes. The misfolded and infectious version is PrP^{Sc}, which is insoluble and believed to contain more beta sheets than PrP^{C}, although the three-dimensional structure is not yet known for certain.

No one knows how PrP^{C} becomes PrP^{Sc}, but Adrian C. Apetri and Witold K. Surewicz at Case Western Reserve University in Cleveland, Ohio, Heinrich Roder at the Fox Chase Cancer Center in Philadelphia, Pennsylvania, and their colleagues studied the folding process of this protein with an experimental technique known as fast mixing. Studying proteins as they fold is exceptionally difficult in the case of proteins that complete the process in tiny fractions of a second. To get the protein to go from a folded to an unfolded state, researchers need to apply solutions with different compositions or pH values. If the introduction of the necessary solution is too slow, the mixing process interferes with the scientists' ability to observe the protein in the act of folding. What is needed is a way to quickly mix the protein and the solution, and then make rapid measurements. Water, the main component of solutions, has a lot of mass and moves slowly, so fast mixing is not easy. Researchers use small containers and special equipment to perform these experiments.

In their experiment, Roder, Surewicz, and their colleagues had the ability to see changes happening on a scale of about 100 microseconds (a microsecond is a millionth of a second), and monitored the protein by certain optical properties that depend on its state. As reported in "Early Intermediate in Human Prion Protein Folding as Evidenced by

Ultrarapid Mixing Experiments," published in a 2006 issue of the *Journal of the American Chemical Society,* what they found was an early intermediate—a transient state—as PrP^{C} folds. The issue now becomes one of determining whether this intermediate is related to the misfolded protein, perhaps by providing an alternate path in which the folding process goes awry. This question can be answered only by further research.

PREDICTING AND MODELING THE FOLDING PROCESS

Scientists continue to develop techniques to watch proteins as they fold, but they also want improved methods to determine the final structure. X-ray crystallography offers excellent resolution but requires a crystal, which is usually not simple to produce and sometimes nearly impossible. NMR spectroscopy offers an alternative to X-ray crystallography, and is performed while the protein is in solution—no crystal is needed. NMR spectroscopy is also a more indirect method, based on advanced physics and the manner in which atoms respond to strong magnetic fields. Unlike X-ray crystallography, NMR spectroscopy provides the positions of hydrogen atoms, but the results are not generally as clear-cut as the X-ray technique, and NMR spectroscopy does not work well with large proteins.

Studying protein structure would be much easier if scientists could tell how the protein folds simply by examining the sequence of amino acids. Bypassing the need for elaborate methods such as X-ray crystallography, and the necessity to generate a crystal, would accelerate protein science tremendously. This seems possible because it is the amino acid sequence that determines a protein's structure. The ability to predict structure from sequence would greatly enhance the understanding of proteins such as PrP^{C} and its misfolded version, PrP^{Sc}, along with mutations and other disease-causing phenomena. Predicting the structure would also be invaluable in efforts to design proteins for certain tasks such as medical treatments. As University of Washington researcher David Baker and his colleagues wrote in *Science* in 2005, "The prediction of protein structure from amino acid sequence is a grand challenge of computational molecular biology."

The goal of much of protein science is to find out how proteins fold. Researchers at the Argonne National Laboratory determined the structure of this protein, called TraR. *(Argonne National Laboratory Media Center)*

But amino acid interactions are hard to predict. At the present time, scientists are unable to deduce a protein's structure from its amino acid sequence, but this topic is the subject of a great deal of research. Databases of known protein structures, such as the Protein Data Bank, are an important resource in this research. Since the amino acid sequences of many proteins are known from genomic studies, researchers often encounter a protein of known sequence but unknown structure. In such a case, scientists search the database for a protein with a similar sequence or at least some portion of similarity. Many proteins share *motifs,* particular structural elements or folds; for example, a leucine zipper is a motif that consists of two alpha helices with leucine amino acids at every seventh position. The two helices join to form a zipperlike structure. Proteins that bind to DNA, such as the proteins that regulate gene expression, sometimes have this motif. If researchers find a new protein with this motif, they have a good idea what kind of functional role it may play in the cell.

Other scientists try to find principles of folding that apply generally to all proteins. This is an extremely difficult task—proteins are large molecules with many types of interaction among their constituent amino acids, as well as with other molecules in their environment (usually an aqueous solution). Researchers who are investigating this issue usually begin with a model, a simplified version of a protein that includes a certain number of amino acids and a certain number of interactions. The model incorporates a large number of properties and data, such as the chemistry of reactions and the physics of atomic forces. To simulate the model's behavior, researchers turn to fast and expensive computers known as supercomputers.

When researchers tackle a complex problem such as protein structure prediction, they prefer to start with the simplest situations. In this case, one of the simplest proteins to study is trp-cage, which contains only 20 amino acids yet has a stable three-dimensional structure and folds very quickly. (Peptides having only a few amino acids form a short chain that does not have enough length to fold.) Trp-cage is a smaller version of a protein found in gila monster saliva, and its structure is known from NMR experiments. Researchers such as Carlos Simmerling, at the State University of New York at Stony Brook, have used a 1,000-processor supercomputer at the Pittsburgh Supercomputer Center to test programs that simulate folding. The programs match the experimental results for trp-cage, so researchers can vary the model by changing an amino acid in the sequence and then make predictions on how the protein will behave.

Computer models that rely only on the sequence of the protein to make predictions of protein structure are called *ab initio,* from the Latin expression meaning "from the beginning." These models do not use any help from experimental data, but start from scratch. There are problems with this approach. In addition to the enormous complexity of proteins, a number of proteins are modified chemically, such as by the attachment of carbohydrates, which may affect their structure. In other cases, chaperones may alter the final state, or restraints caused by location or interaction with other molecules could influence the protein's structure. These possibilities complicate the prediction structure from the sequence because they are not always reflected in the sequence—the amino acids generally contain all the information needed for a protein to fold, but certain circumstances may put a kink or two in the chain.

Further progress will come with more tests of the models as well as faster computers to simulate them. Toward this goal, computer company IBM has designed an exceptionally fast computer architecture known as Blue Gene, specifically with biological applications in mind. The computer Blue Gene/L, which uses this architecture, is one of the fastest computers in the world.

Other projects enlist help from computer users all over the globe. Projects such as Rosetta@home and Folding@home invite computer owners to participate in scientific research by downloading a program onto their home computer. This program will run when the owner is not using the computer, in which case the computer gets linked to a network that conducts simulations and solves problems related to protein structures. Such distributed computing makes good use of computer resources that would otherwise be wasted. The team at Stanford University in California that operates Folding@home has made numerous investigations based on these simulations, including research into how proteins fold in tight spaces. The operator of Rosetta@home, David Baker at the University of Washington, has used his distributed computing resources in numerous projects, such as collaborations with scientists who are studying the proteins in anthrax toxin. See the Further Resources section for further information and Internet addresses for Rosetta@home and Folding@home.

PROTEIN DESIGN

As shape and folding are better understood, researchers can begin to think about how to tinker with proteins—and even design a few of their own. Nature tinkers with proteins by introducing mutations in genes that code for proteins, resulting in slightly different amino acid sequences. Most of these mutations make the protein less effective and are weeded out by evolution, but sometimes the protein becomes more efficient or serves a new, advantageous function. With an increased knowledge of protein structure, people could begin to design proteins to suit a variety of purposes in medicine and engineering.

In 2003, Brian Kuhlman, David Baker, and their colleagues designed a novel protein. This protein, which contains 93 amino acids and was called Top7, is completely new—it is not found in nature. As reported in *Science* in "Design of a Novel Globular Protein Fold with Atomic-Level Accuracy," the researchers designed the protein based on a structure they wished to create, made the protein using enzymes, and then studied it with NMR spectroscopy and X-ray crystallography. The structure matched the prediction.

This startling experiment demonstrated the progress scientists have been making in the study of proteins and their structure. But it is only a first step. Baker and his team did not design Top7 to have any particular function; they simply proved that in at least one case they could successfully design a protein by predicting which sequence of amino acids would produce a simple structure.

To create proteins that meet specific requirements, researchers need a better understanding of how proteins work. Scientists must learn how to predict a large number of structures, including complicated ones, and discover how these structures enable proteins to carry out their important functions. This knowledge would allow researchers to tweak a protein by substituting an amino acid or two, or make other slight changes that would enhance the protein's existing function or create a protein that does something entirely new. Nature does this slowly using random mutations, but nature has virtually unlimited resources and is not on a tight schedule. People need to work faster and smarter in order to direct the process toward specific goals.

In 2004, Mary A. Dwyer, Loren L. Looger, and Homme W. Hellinga of Duke University in North Carolina achieved some success in changing a protein that normally has no catalytic activity. As reported in the *Science* paper "Computational Design of a Biologically Active Enzyme," the researchers made 18–22 mutations, based on their computer models, to give the protein the ability to catalyze certain reactions. In some cases, the altered protein increased the rate of the reaction up to a million times.

Adjusting or adapting a protein for a specific purpose has extremely important medical applications. Proteins are active molecules that perform a wide variety of functions, giving them a great deal of potential to cure diseases. But in any medication, unwanted side effects can overshadow or offset the benefits. Side effects often occur when drugs act on healthy tissue in addition to their intended target, so drug manufacturers strive to make their products as specific as possible by various chemical modifications, which are often tough to control. Proteins are easily adjustable, and if researchers can get a handle on how to design the appropriate changes, they can attain specificity and avoid most side effects. In addition to acting on diseased tissue, novel or adapted proteins may also be able to deliver medicine, monitor the body for any signs of disease, and perform other jobs to maintain health.

CONCLUSION

The study of protein structure and function received a tremendous boost from the completion of genome projects such as the Human Genome Project, which among other contributions provided the sequences of proteins made by the human body. Painstaking methods for determining protein structures have accumulated much data, but researchers still have a way to go before they completely understand how sequences of amino acids specify a protein's structure, and how to adapt or create proteins for important new tasks. As Mary A. Dwyer and her colleagues wrote in their 2004 *Science* paper, designing proteins "has tremendous practical potential . . . but presents a formidable challenge and is one of the most stringent tests for understanding protein chemistry."

As cellular machines, proteins are also of great interest to researchers working in the field of nanotechnology. These researchers aim to design and construct tools and devices on a nanometer scale—100,000 times smaller than the width of a human hair. With their tiny size and functionality, protein molecules are good candidates for this technology. Proteins such as kinesin and dynein offer an example of how tiny motors can work.

Protein motors have been used by researchers to transport such objects as tiny particles, nanowires, bits of polystyrene and glass, and DNA, and have played roles in creating small rotating propellers and pumps that drive fluids through hair-thin pores. In 2006, Peter Michaely at the University of Texas Southwestern Medical Center in Dallas, Vann Bennet and Piotr E. Marszalek at Duke University, and their colleagues discovered that a common protein motif called ankyrin repeats would make an excellent spring. Using atomic force microscopy, which can measure the properties of single molecules, the researchers found that ankyrin can stretch and snap back into place, generating a force. These molecules behave as linear springs—the force they produce varies directly in proportion to the distance they are stretched—unlike other proteins, which behave in more complicated ways when stretched and compressed. This means they would be an ideal choice for a nanotechnology engineer who needed the molecular equivalent of a coiled spring, which is an exceptionally useful device—larger versions of coiled springs find use in everything from mattresses to automobile suspension systems.

Advances in the study of protein structure and function will impact science, engineering, and medicine in many ways. As this field of research grows, so will the ability to harness these versatile cellular machines. Making proteins is not difficult—for instance, bacteria "factories" are available, as are other systems—but designing the right protein for the right job is presently a hard problem. This research is in its early stages, but it is a frontier that has plenty of room to grow.

CHRONOLOGY

1839 C.E. Dutch chemist Gerardus Johannes Mulder (1802–80) isolates and analyzes a substance from plants and animals that he calls protein, from the Greek *prōteios,* primary or first.

1895 German physicist Wilhelm Röntgen (1845–1923) discovers X-rays.

1899–1908 German chemist Emil Fischer (1852–1919) shows that proteins are composed of amino acids.

1912–14 German physicist Max von Laue (1879–1960) and British scientists Sir William Henry Bragg (1862–1942) and his son, Sir William Lawrence Bragg (1890–1971), discover how to use X-ray diffraction to analyze the structure of molecules.

1950 American chemist Linus Pauling (1901–94) deduces the structures of alpha helices and beta sheets in proteins.

1951–55 British chemist Frederick Sanger (1918–) discovers the sequence of amino acids in the protein insulin, the first sequence to be fully determined.

1958 British biochemist Sir John Cowdery Kendrew (1917–97) and his colleagues determine the first protein structure (myoglobin) by using X-ray analysis.

1962 Based on a series of experiments, American chemist Christian Anfinsen proposes that proteins fold spontaneously into a structure representing the most stable state.

1968 Because proteins fold quickly but have a large number of possible conformations, American biologist Cyrus Levinthal proposes that proteins must reach their folded state in a direct manner.

1971 Initially containing seven structures, the database known as Protein Data Bank begins at Brookhaven National Laboratory.

1978 Researchers at the City of Hope National Medical Center in California and the biotechnology company Genentech insert the gene to make human insulin into bacteria, and coax the bacteria into producing large amounts of the protein.

1982 American biologist Stanley Prusiner discovers prions.

1985 Colin L. Masters of the University of Western Australia and his colleagues at Royal Perth Hospital in Australia and the University of Cologne in Germany find misfolded amyloid protein in the plaques associated with Alzheimer's disease.

1998 Rosetta@home, a distributed computing project involving many home computers, launches.

2000 Folding@home, a distributed computer project involving many home computers, launches.

2003 The Human Genome Project is completed.

American scientist David Baker and colleagues design a novel protein, Top7, based on computational models and the principles of chemistry.

2004 IBM constructs Blue Gene/L, the first supercomputer with the Blue Gene design, which is specifically geared toward solving computational problems in the biological sciences.

2008 The National Cancer Institute, a division of the National Institutes of Health, sponsors a conference in Amsterdam for the international scientific community to discuss policies governing the public release and sharing of proteomics data.

2009 The Human Proteome Organisation, an international organization devoted to proteomics, holds its 8th Annual World Congress in Toronto, Canada.

FURTHER RESOURCES

Print and Internet

Apetri, Adrian C., Kosuke Maki, Heinrich Roder, and Witold K. Surewicz. "Early Intermediate in Human Prion Protein Folding as Evidenced by Ultrarapid Mixing Experiments." *Journal of the American Chemical Society* 128 (2006): 11,673–11,678. The researchers found an early intermediate—a transient state—as the human prion protein folds.

Argonne National Laboratory. "Structural Biology Center." Available online. URL: http://www.sbc.anl.gov/. Accessed April 1, 2009. This Web resource offers a public tour of the Argonne National Laboratory facilities, a gallery of protein structures, and highlights of recent research.

Baker, David. "Rosetta@home." Available online. URL: http://boinc.bakerlab.org/rosetta/. Accessed April 1, 2009. This Web site explains how computer owners can participate in protein research at home by downloading Rosetta@home software.

Clark, David P., and Lonnie D. Russell. *Molecular Biology Made Simple and Fun,* 3rd ed. St. Louis, Mo.: Cache River Press, 2005. This highly enjoyable and accessible book offers much information on DNA and proteins.

Dwyer, Mary A., Loren L. Looger, and Homme W. Hellinga. "Computational Design of a Biologically Active Enzyme." *Science* 304 (June 25, 2004): 1,967–1,971. The researchers mutated a protein to give it the ability to catalyze specific reactions.

Kuhlman, Brian, Gautam Dantas, Gregory C. Ireton, Gabriele Varani, Barry L. Stoddard, and David Baker. "Design of a Novel Globular Protein Fold with Atomic-Level Accuracy." *Science* 302 (November 21, 2003): 1,364–1,368. The researchers report the design of a new protein.

Martz, Eric, and Trevor D. Kramer. "World Index of Molecular Visualization Resources." Available online. URL: http://molvis.sdsc.edu/visres/index.html. Accessed April 1, 2009. This Web page provides a huge list of links to galleries, tutorials, and animations on the structure of molecules.

National Institute of General Medical Sciences. "Protein Structure Initiative." Available online. URL: http://www.nigms.nih.gov/Initiatives/PSI/. Accessed April 1, 2009. The Protein Structure Initiative, sponsored by the National Institute of General Medical Sciences, a branch of NIH, provides support to researchers involved in determining the three-dimensional structure of proteins. The Web page presents examples of these structures along with news and information.

Pande, Vijay. "Folding@home." Available online. URL: http://folding.stanford.edu/. Accessed April 1, 2009. Similar to Rosetta@home, participants in Folding@home can put their computer downtime to good use, advancing protein science.

Peters-Libeu, Clare A., Yvonne Newhouse, Danny M. Hatters, and Karl H. Weisgraber. "Model of Biologically Active Apolipoprotein E Bound to Dipalmitoylphosphatidylcholine." *The Journal of Biological Chemistry* 281 (2006): 1,073–1,079. The researchers found the structure of an important protein called apolipoprotein E (abbreviated apoE) while it was bound to a lipoprotein (a molecule containing fats and protein).

Rensberger, Boyce. *Life Itself: Exploring the Realm of the Living Cell.* Oxford: Oxford University Press, 1998. The author, a skilled science writer, discusses the structure and molecules of the cell, including proteins.

Uversky, Vladimir N., Alexander V. Kabanov, and Yuri L. Lyubchenko. "Nanotools for Megaproblems: Probing Protein Misfolding Diseases Using Nanomedicine *Modus Operandi.*" *Journal of Proteome Research* 5 (2006): 2,505–2,522. The researchers review protein misfolding diseases and how scientists are using nanotechnology—technology on a small scale—to tackle the problems.

Web Sites

Fox Chase Cancer Center. Available online. URL: http://www.fccc.edu/. Accessed April 1, 2009. Readers can learn more about the Fox Chase Cancer Center at this Web site, which offers information on its research programs as well as important cancer information and news.

Pittsburgh Supercomputer Center. Available online. URL: http://www.psc.edu/. Accessed April 1, 2009. Operated by the University of Pittsburgh, Carnegie Mellon University, and Westinghouse Electric Company, the Pittsburgh Supercomputer Center maintains a number of powerful computers. Researchers from many different universities and institutions have used this facility to study a variety of scientific problems, some of which involve protein structure and function, as described in the research and news pages of this Web site.

Protein Data Bank. Available online. URL: http://www.rcsb.org/pdb/home/home.do. Accessed April 1, 2009. The large database of protein structures along with some RNA structures contained in the Protein Data Bank is accessible through this Web site.

4

Biodiversity—The Complexity of Life

More than half of the world's almonds come from California. Production of these nutritious nuts—the seeds of the fruit of almond trees—is a billion-dollar industry in this state, in which the mild winters and extensive, dry summers provide the environment that almond trees need to grow. Originally from the Mediterranean region, almond trees made their appearance in California when Spanish settlers transported them in the 18th century. Yet despite the advantages of California's climate and the skill of its agriculturists, production would be worthless without the help of an insect—the bee.

Almond trees cannot self-fertilize, so they must be cross-pollinated—pollen must be transferred from one tree to another. Bees do this as they fly from blossom to blossom in search of nectar. California farmers have so many almond trees needing pollination that they must hire beekeepers to place hives near their land. In late February and early March, beekeepers in California and other states truck their honeybee hives into position for the almond season. Other crops such as alfalfa, apples, cherries, and plums also use bee pollination. The price beekeepers charge per colony held steady from 1995 through 2004 at around \$40 to \$50, but in 2005 the price was much higher, and in 2006 it hit nearly \$140 before stabilizing to its current price of about \$150 as of 2009. The problem was that disease was wiping out the honeybee population. According to the United States Department of Agriculture, insect-pollinated plants

Bees perform valuable, although unwitting, pollination services. *(Skynesher/Dreamstime.com)*

account for about one-third of human food sources. Bee populations continue to be low, and the lack of these prolific pollinators is troubling.

The link between crops, food production, and the health and welfare of bees is an example of the interdependence of life forms on Earth. A disruption or decimation in one species affects others, which in turn affect still others. Although biologists have spent a lot of time cataloging species, the interactions among these organisms and with their environment are complex and not well understood. This chapter describes research that explores biological diversity—*biodiversity*—and how the diversity of organisms affects an environment's ability to maintain itself and produce food. As the bee situation clearly demonstrates, biological interactions have extremely important ramifications for the environment as well as the pocketbook.

INTRODUCTION

Swedish botanist Carl von Linné (1707–78), who published books and papers under Carolus Linnaeus, his Latinized name, was one of the first people to attempt to order and classify the wide variety of life on Earth. In 1735, Linnaeus published the first edition of *Systema Naturae* (Latin for "system of nature"), in which he divided nature into the plant, animal, and mineral kingdoms, and placed species in various classes, orders, and genera based on similarity. Linnaeus went on to publish 13 editions of this book, classifying well over 10,000 plants and animals. The figure shows one example of a classification.

Since that time, biologists have discovered many more organisms, including a vast array of single-celled microorganisms. About

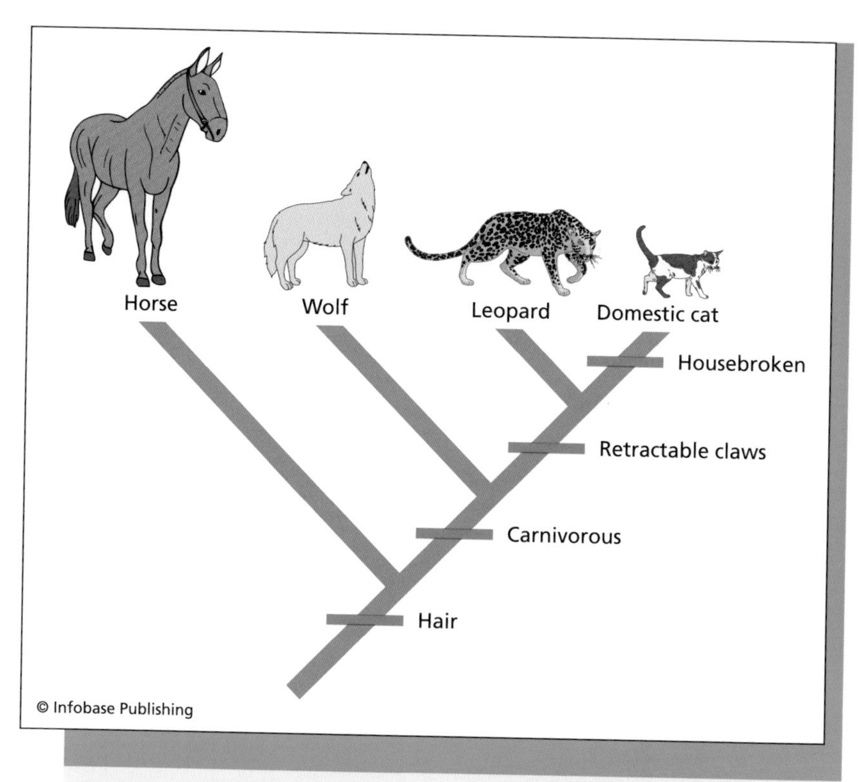

Animals can be distinguished from one another on the basis of observable traits or behaviors, such as the presence of hair, as in mammals, a diet of meat, as in carnivores, and so on, for each species.

Coral reefs harbor many types of organisms. *(Kopstal/Dreamstime.com)*

1.75 million species have been identified, and more are being added to the list each year. For example, New Zealand scientist Brian Smith and his colleagues discovered a new species of caddisfly caught in a trap in the city of Hamilton, New Zealand, in 2007—an unexpected result, since most new species are found in rural areas instead of cities. Tropical regions such as the South American rain forests surrounding the Amazon River are as yet poorly explored and undoubtedly harbor many more species; no one knows how many organisms still await discovery, but estimates for the total number of species on Earth range from 5–100 million.

Biologists do not always agree on the definition of the term *species.* The old definition is based on reproduction—a species is a group of animals that can interbreed. In this age of DNA sequencing and genomics, another way of measuring relatedness of organisms is to compare their genetic information. But either way it is defined, the notion of species, and the large number of them inhabiting the planet, adequately conveys the biological diversity of life on Earth. The term *diversity* refers to groups that are composed of different elements or individuals; biological diversity is often shortened to biodiversity.

There are various ways of defining biodiversity, as there are for defining species, since diversity can be considered in terms of organisms or in terms of genes and genetic information. Scientists can also consider the distribution of this variety. Some regions have more diversity than others; for example, the tropics—near the equator—harbor more different types of organisms than the North and South Poles.

Why is life on Earth so diverse? This is not an easy question, and scientists are uncertain of the answer. One part of the answer must involve evolution, since fossil evidence shows that life arose at least three billion years ago, and the species that exist today are the product of subsequent evolution. The principles of evolution include variability in genetic information (such as mutations), inheritance of traits, and natural selection—the process by which organisms that are better adapted to their environment are able to reproduce, and the poorly adapted organisms vanish. Changes in an organism's shape and properties slowly arise because of evolution, resulting in the wide variety of life existing today. But why there is so much biodiversity, and how the planet supports this rich variety of life, are issues still under debate. Some of this debate involves research activities that will be discussed in this chapter.

Although the underlying reasons for diversity may not be clear, the consequences of a lack of diversity are definite. Catastrophes such as the potato blight of the 1840s in Ireland, which caused a severe famine due to the country's reliance on this crop, are an example of the old proverb that warns against putting all one's eggs in the same basket—a single accident can ruin the whole lot.

But diversity has many other benefits. Ecology is the study of the relationship between organisms and their environment, and an ecological system—*ecosystem* for short—is a community or habitat of organisms interacting with each other and with the environment. Many people enjoy visiting ecosystems such as those preserved in parks and national wildlife reserves, but ecosystems offer much more than that. The list of benefits provided by Earth's ecosystems includes an abundance of game animals and fish, important supplies of fresh water, the maintenance of soil fertility, and the decomposition of waste materials. These services are often taken for granted, and other benefits, such as the large number of medications derived from plants, are not widely known or appreciated. Although many drugs these days are made in a laboratory, at least 25 percent of prescribed medications include active ingredients extracted from plants. For instance, digitalis medications such as digoxin, commonly prescribed for certain heart conditions, come from *Digitalis* plants (also known as foxglove). Quinine, once the most popular drug to treat malaria and still occasionally used, comes from *Cinchona* trees.

In the 1970s and '80s, many people believed that diversity was the result of a productive and stable environment. In this view, diversity was simply a byproduct of an ecosystem. More recently, though, scientists are realizing that the situation is much more complex. Researchers have found evidence that diversity may not be just a consequence of a rich and long-lasting ecosystem, but instead may actually contribute to the ecosystem's productivity and stability. In other words, diversity is important in ecosystem function, and a loss of diversity may seriously impact the production and stability of an ecosystem, along with the benefits humans reap from them. This research issue is the main thrust of this chapter.

There are several reasons why this research is important, but one reason has gained a lot of attention in recent years—Earth is currently experiencing a well documented loss of biodiversity. Species are disappearing, a process called extinction. Extinctions are nothing new, and the fossil record indicates that they have occurred continuously throughout Earth's history. It is normally a slow process, although in rare cases, a lot of extinctions occur at roughly the same time, and the planet has suffered at least five major extinction events in which a large percentage of species vanished. These events had devastating consequences. The most recent took place about 65 million years ago—famous for killing off the dinosaurs—and may have been caused by a meteorite or comet strike.

According to the World Conservation Union, an international organization that promotes resource conservation, more than 800 species have become extinct in the last 500 years. This rate is many times higher than what the fossil record indicates is typical. The losses will probably continue, as there are 16,928 threatened species on the 2008 World Conservation Union's Red List. Human activity is almost certainly responsible for at least some of these losses. Industrial pollution, along with commercial and residential development, has destroyed many habitats, and excessive hunting and fishing has dangerously thinned a number of animal populations. Some species have also been crowded out of their habitats by nonnative "invasive" species, accidentally or sometimes intentionally introduced into new habitats by humans. The extinctions are so widespread that, according to *Global Biodiversity Outlook 2,* a 2006 report issued by the Convention on Biological Diversity, "In effect, we [humans] are currently responsible for the sixth major extinction event in the history of the Earth. . . ."

No one is yet certain what effects this continuing loss of biodiversity will have on the health and welfare of Earth's ecosystems, including those on which humans rely for food and industry. This is a question for science to answer.

COEXISTENCE—LIVING AND INTERACTING TOGETHER

The number of species in a specific environment is an important property of that environment, but the species themselves are perhaps less important than the way they interact with one another and their surroundings. Many different levels of interaction are possible, even with animals that do not directly encounter one another yet need the same resources—food, water, and shelter. As the number of different species in an ecosystem rises, the complexity of those interactions becomes enormous and makes it difficult to understand how the ecosystem operates. This is part of the complexity of life on Earth.

Scientific research often proceeds by breaking down complex problems into simpler elements, a process known as analysis. (The term *analysis* comes from a Greek word *analyein,* meaning to break up or loosen.) The different kinds of organism interactions can be broken down into groups, some of the most important of which are predation, mutualism, commensalism, and competition.

Predator and prey relationships get a lot of attention on wildlife shows on television. This relationship can be quite dramatic, such as a pride of lions hunting wildebeest in southern Africa. But the same category in its broadest sense includes all levels of *trophic* interactions (trophic refers to nutrition and eating). A cow grazing on grass in a pasture is not as suspenseful as a lion hunt but is also an interaction

A prominent interaction among organisms involves a predator—in this case, the bird—and prey (the insect). *(Cay-Uwe Kulzer/iStockphoto)*

involving nutrition. Food is one of the basic necessities for all forms of life, and all organisms need energy for growth and maintenance. One organism may feed on plants, and this organism may in turn provide food to another, which may be eaten by yet another; these interactions create lengthy food chains. The chains may branch, with some species participating in multiple chains, which creates complex interactions known as *food webs.*

Another category of interaction is mutualism, in which both participants mutually benefit. For example, in the search for nectar, bees travel from plant to plant and in the process transfer pollen, which is a powdery substance containing reproductive cells of the plant. Instead of releasing their pollen and relying on random winds to blow some of it to the target, some plants attract insects with their nectar. As the insects collect the nectar, some pollen rubs on and off their bodies and is transferred from plant to plant. The insects benefit from the nectar, and the plants benefit from the pollen transfer. Insect-pollinated plants account for about one-third of human food sources, and honey bees do the majority of the work.

Commensalism is an interaction in which only one species benefits, similar to predation except in commensalism the other species is not affected. (The term *commensalism* derives from Latin words *com,* meaning sharing or together, and *mensa,* meaning table.) For example, birds called egrets eat insects stirred up by the cattle as they graze.

Competition occurs when species compete for resources. This is an interaction that generally does not directly benefit either species. Consider two corporations engaging in a price war to win business; both may suffer a temporary reduction in profits. Competition between animals for the same prey or the same shelter usually means there is less available for each.

Sometimes a single species in an ecosystem has such a significant impact that it dominates the interactions. The removal of this species, known as a keystone species, would drastically alter the ecosystem. (The word *keystone* refers to a vital component, as in the keystone of an arch that locks the other pieces into place.) Examples include beavers, which are a keystone species for certain ecosystems because of the importance of the dams they build; elephants, whose browsing prevents grasslands from becoming overrun with trees; and certain starfish that are the sole predators of a number of marine species. Take away these animals, and their ecosystem would become much different.

But all species in an ecosystem contribute to the whole, and the ecosystem is the sum of a large number of interactions. Biodiversity is important because different species have different traits and different strengths and weaknesses. If a single species was optimally suited for all aspects of its environment—for instance, if it could take equal advantage of any food source or shelter—then it should be the only species in its ecosystem. If this was the case, there would be one species that lived in deserts, one in tropical climates, one in tidal coastlands, and so on. But instead, each ecosystem hosts a variety of organisms because species tend to specialize, such as bees making honey and lions chasing wildebeest. Organisms exploit their strengths, and different species coexist by exploiting different strengths.

Research into the influence of biodiversity on ecosystems has focused on two ecosystem properties, stability and productivity. In general, stability refers to constancy and the ability to maintain a steady level or amount. Ecosystems that are not stable tend to vary in terms of the number and perhaps composition of species, and may require a long time to return to normal if something such as a drought, an invasive species, or a disease outbreak upsets the balance. Instability can cause severe disruptions in a number of ways, such as a sudden increase in locusts or mice that destroy grain crops and cause food prices to surge or, in the worst case, create famine. Productivity, on the other hand, is a measure of abundance. Scientists often determine productivity in terms of *biomass*, the weight of organisms. (In practice, such measurements are usually performed on plant species—plants can be cut and weighed.) Productivity is a measure of quantity, while stability is a measure of variations over time.

COMMUNITY STABILITY AND PRODUCTIVITY

An early study on the relationship between biodiversity and stability involved the theoretical work of Australian-British scientist Sir Robert May (1936–   ) in the 1970s. Before this work, some scientists such as British biologist Charles Elton (1900–91) proposed in the 1950s that biodiversity influenced an ecosystem. Elton reasoned that simple ecosystems, which contain little diversity, would be more subject to fluctuations than a highly diverse ecosystem, and therefore have less stability. Although this assertion sounds reasonable, there was little experimental or theoretical evidence for it. The idea was temporarily abandoned in the 1970s and 1980s, mostly due to May's work.

May, who was trained in physics and mathematics, tackled biological problems with a rigorous mathematical approach. He constructed mathematical models that abstracted and quantified features of ecosystems, and developed equations predicting how the system evolves over time. May's models showed that higher biodiversity tends to make individual species less stable, which means their populations will vary considerably over time.

Modern researchers still believe May's models are correct, but there is another issue to be considered. If each species is less stable, then is the ecosystem as a whole less stable? It might seem reasonable to assume this is true; if so, then Elton's hypothesis is highly unlikely, since higher diversity should result in less overall stability because the member species are less stable. As David Tilman, a researcher at the University of Minnesota, said in a 2001 interview posted at In-cites.com, "The problem was that many ecologists assumed in the 1970s that, if each individual species was less stable at higher diversity, then the sum of these species, the abundance of the whole ecosystem, would also be less stable." But the research of Tilman and others has suggested otherwise.

Tilman and his colleagues started maintaining plots of grassland in the Cedar Creek Natural History Area in Minnesota in 1982. The plots were square and about the size of a room in a house, about 13.1 feet (4 m) per side. To make biomass measurements, Tilman's team cut the plants in a narrow strip and sorted the species, then measured the mass of each. This strip was a sample—cutting and sorting the whole plot would take too long, so the researchers took a sample and estimated the entire biomass from this measurement. Estimations based on samples involve a branch of mathematics called statistics, which is used a great deal in science and engineering.

A drought occurred in 1988. The drought decreased biomass considerably, which is not surprising, but over the next few years Tilman noticed something interesting—the plots that had more species fared better than those had few. Although the masses of individual species varied, plots with high biodiversity had a more stable biomass overall than plots with low biodiversity.

The drought data seemed to show a correlation between biodiversity and stability—the higher the biodiversity, the greater the stability, which is a different conclusion than May had come to in the 1970s. But Tilman's observation by itself was not convincing because there was no control over the other variables; researchers performing a laboratory experiment try to control all variables except one, in which case any differences in the measurements must be due to this variable. In Tilman's data, the researchers had made no attempt to control any of the variables. Biodiversity was one variable, but there are others that could influence biomass, such as soil quality and temperature.

To study the relationship between biodiversity and stability, Tilman and John Downing, a colleague at Iowa State University, used statistics. With the aid of advanced statistics, they found that biodiversity seemed to be the most important variable. Statistics deals with probabilities, not certainties, but the results showed that biodiversity was far more likely to be the main factor associated with stability. Tilman and Downing wrote their results in an influential paper, "Biodiversity and Stability in Grasslands," published in *Nature* in 1994.

This paper touched off a heated debate among ecologists. Critics pointed out that there could have been other variables that Tilman and his colleagues did not measure. In response to this valid criticism, Tilman and his team, and other scientists, set up controlled experiments. Such experiments in the field are difficult, requiring a lot of work to maintain the plots and keep variables under control. (As most gardeners know, plots of land, including back yards, tend to get weedy and overgrown without careful attention.) The results of these experiments also indicated a direct relationship between biodiversity and stability.

Similar experiments by other researchers include a large project called BIODEPTH, which stands for Biodiversity and Ecological Processes in Terrestrial Herbaceous Ecosystems. Beginning two years after the paper by Tilman and Downing, these studies also established controlled grassland plots, but did so at sites that differ greatly in soil and climate—Britain, Ireland, Portugal, Germany, Switzerland, Greece, and Sweden.

BIODEPTH experiments, along with those of Tilman's group, indicated an important role for biodiversity, which was also linked with productivity. Plots containing half as many plant species as another plot are 10–20 percent less productive. If a plot is filled with only one plant species, it will prove to be only half as productive as a plot that harbors a variety of 24–32 species.

The studies on ecosystem stability and productivity suggest that these properties and biodiversity are correlated—when biodiversity increases, so does stability and productivity. But there remain some questions about the nature of this correlation. A correlation between two variables does not necessarily mean that one causes the other, because it is possible that a third, hidden variable is actually causing both of the other two variables to increase or decrease at the same time. In this case, a hidden variable could be increasing both biodiversity and stability, creating the correlation. Even with controlled experiments researchers sometimes have difficulty establishing a cause-and-effect relationship between variables, because the researchers may not be certain that all the potentially influential variables have been controlled.

To establish the nature of the relationship between ecosystem properties and biodiversity, researchers need to explore the ways in which biodiversity can influence these properties, should this prove to be the case. In other words, researchers must seek the mechanisms by which biodiversity can affect stability and productivity.

As for stability, three hypotheses have emerged. A diverse ecosystem may be more resistant to change because of an average effect—in statistics, for example, when one averages a large number of random variables, the results tend to be less variable than when one averages a smaller number. Another factor comes into play because different species have different strengths but compete for the same resources, and therefore their abundance tends to move in opposite directions—as one increases, the other decreases. When all of these interactions are summed, as they are in an ecosystem, the result is a diminishing of overall variability. Yet another factor is known as the insurance effect: A greater number of species means there is a greater chance that one or more species will thrive as environmental conditions vary. This results in a decrease in variability, as there will probably be at least one species that will make up any deficits created in the new environment and ensure the community continues to thrive. All of these explanations may be important.

Scientists have also considered several hypotheses to explain the relationship between biodiversity and productivity. One possible explanation for the greater productivity of the diverse plots is that different plants may have different strengths, with each taking advantage of the resources in a different way. As a result, the resources of the environment are more fully utilized, and a plot with a given set of resources will be more productive than when hosting only a few species.

Another potential explanation for the increased productivity is particularly important because it illustrates the possibility that there is another variable besides biodiversity to be considered. Suppose there is a certain species that is ideally suited to the environment. This species would be extremely productive, and its presence would strongly affect an ecosystem's productivity. The effect would be similar to that of a keystone species. Because an experimental plot with more species has a better chance of containing this species, it will also have a better chance of being more productive. In this view, the cause of the increase in productivity is not due to biodiversity itself, but rather to one or perhaps a few essential species. Biodiversity in this case would not matter; only the presence of the most productive species would be critical, and biodiversity seems to play a role only because it gives the plots a better chance of containing the MVPs (most valuable players).

Tilman and his colleagues Peter B. Reich, Troy Mielke, and Clarence Lehman, also of the University of Minnesota, and Johannes Knops and David Wedin of the University of Nebraska, addressed this issue in a 2001 paper, "Diversity and Productivity in a Long-term Grassland Experiment," published in *Science*. Using the same kind of techniques described earlier, the researchers investigated not only species number but also species composition. They found that diverse plots performed better than even the most productive plot consisting of a single species; this result suggests that biodiversity was the critical factor rather than which species happened to end up in the plot.

Field experiments help scientists learn more about biodiversity issues by simplifying the environment, which enables the scientists to control various properties. The simple environments permit researchers to draw meaningful conclusions. But the world is a more complicated place than experimental plots, and a question always lurks when scientists review these experiments: Do the results apply to larger, more complex ecosystems? Answering this question is difficult because of the presence of many variables and a lack of control in these larger systems.

There is still debate among ecologists, and as yet no consensus has been reached on this issue. Some scientists, such as Michael Huston at Texas State University, have strongly criticized the biodiversity experiments described above, claiming that the results may not be valid in naturally occurring ecosystems. Huston pointed out in a letter published in the January 24, 2003, issue of *Science,* "many of Earth's highest diversity areas have low productivity," such as certain shrub lands of South Africa and Australia. The debate is rancorous partly because of the scientific disagreements, but also because the findings of ecological research are often used to guide a government's policies concerning the environment.

FOOD CHAIN ANALYSIS

By focusing on a specific set of important interactions, some researchers hope to gain a better understanding of how complex ecosystems work, and obtain decisive answers to the debates on biodiversity. Some of the most important interactions involve the food webs mentioned earlier.

Experiments can limit the interactions of an isolated plot to a certain minimal level, perhaps only a single link in the food chain—a herbivore species (plant-eaters) and a plant species. The more complex environments that exist in nature consist of multiple levels, with a variety of plants, herbivores, and carnivores (meat-eaters). Studying these interactions relies on knowing what species live in the ecosystem and who interacts with whom. In the case of food chain studies, this means who eats whom.

One way to determine an animal's diet is by watching it in its natural environment. But such field observations take a lot of time, and can be nearly impossible for shy animals and those that live in hard-to-see places such as lakes or rivers. Researchers can catch a sample of these animals and analyze the contents of the digestive system, but it is difficult to make precise determinations with these data.

Researchers with a more technological orientation have another tool—stable *isotope* analysis, as discussed in the following sidebar. Isotope studies can determine a particular animal's position in a food chain by measuring how far it is from the beginning (the primary food producer, often a species of plant). For example, M. Jake Vander Zanden and Joseph B. Rasmussen of McGill University in Montreal, Canada, Brian J. Shuter of the University of Toronto in Canada, and Nigel Lester at the Ontario Ministry of Natural Resources in Canada used the isotope nitrogen-15 to study 14 lakes in Ontario and Quebec, and published their report, "Patterns of Food Chain Length in Lakes: A Stable Isotope Study," in *The American Naturalist* in 1999. At the

Food Chains and Isotopes

A chemical element has a specified number of positively charged particles called protons in its nucleus, which is the element's atomic number and represents the element's place in the Periodic Table of Chemical Elements. But there are a variable number of electrically neutral particles called neutrons in the nucleus, and different isotopes of a given element have a different number of neutrons. For instance, carbon-12 has six protons and six neutrons (for a total of 12 particles), and carbon-14 has six protons and eight neutrons. Some isotopes are radioactive, meaning they emit radiation and decay into other isotopes; cobalt-60 is an example of a radioactive nucleus, as is carbon-14, which is used for radiocarbon dating (determining the age of biological samples). Other isotopes are stable and do not undergo decay. Nitrogen-15 is a stable isotope, as is nitrogen-14, which is the most common isotope of nitrogen (more than 99 percent of this element on Earth is nitrogen-14).

top of the food chain in all the lakes was lake trout, but the lakes had a variety of trophic levels—in some cases the lake trout were only two levels above the lowest, indicating that the food chain was short, and in other cases the lake trout were four levels higher.

Another potential tool to study food webs is DNA analysis. Organisms have a species-specific set of genes that contain the instructions for the development and maintenance of all their cells and tissues. Scientists have recently sequenced all the genes—the genome—of a number of organisms, and are continuing to sequence the genome of more organisms. Knowing the DNA sequence of a certain species allows scientists to identify a member of that species based on its DNA, sometimes by just a small part of this DNA. With the help of DNA technology,

At the base of a food chain is a producer such as a plant, which makes carbohydrates with the energy of sunlight in a process known as photosynthesis. Animals cannot do this, so they rely on plants or other animals for food. The next step in the food chain, for example, might be a small herbivore. The next highest level could be a small carnivore, which in turn is prey for a larger predator. At each step in this chain, the stable isotope nitrogen-15 accumulates because animals tend to retain nitrogen-15 more than nitrogen-14. (Researchers are unsure why this happens.)

Larger animals have more nitrogen-15 than smaller animals simply because of their bulk, not necessarily because of their position in the food chain. Scientists therefore measure the ratio of nitrogen-15 to nitrogen-14; the value of this ratio does not depend on the animal's size, and increases by 3–4 percent with each level in the chain. For instance, people who eat only vegetables—vegetarians—have a lower nitrogen-15/nitrogen-14 ratio than meat-eaters. To make the measurement, researchers often use instruments called mass spectrometers to separate the isotopes based on the small differences in their mass.

analyzing the contents of an animal's digestive system will soon become much easier and more precise.

Researchers are still far from being able to study complex ecosystems and food webs with as much thoroughness and precision as they would like. But as the tools and techniques of ecosystem analysis improve, the interactions that govern its properties will become clearer.

Meanwhile, some researchers are studying complex ecosystems by recreating their rich interactions in the laboratory. One of the biggest projects is Biosphere 2, a 3.14-acre enclosed structure in Oracle, Arizona. Finished in 1991, Biosphere 2 is 91 feet (27.7 m) at its highest point and contains 6,500 windows and about 7,000,000 cubic feet (200,000 cu. m) of space. The ambitiousness of the structure was reflected in its name—the biosphere is the portion of Earth in which life exists, and Biosphere 2 was meant to be a recreation of the biosphere on a smaller, more manageable scale.

A company called Space Biosphere Ventures constructed Biosphere 2, and used it to investigate the possibilities of enclosed habitats in space or colonies on other worlds. Since then, ownership and management has changed hands several times, and in June 2007, the University of Arizona announced it would manage the facility. Biosphere 2 contains a variety of habitats, including a miniature savannah, mangrove, desert, and a million-gallon (263,000 l) tropical sea. Researchers will use Biosphere 2 to study a variety of ecological questions, and the University of Arizona intends to do research on climate and water issues.

Biosphere 2 offers the advantage of bringing a number of complex ecosystems into the laboratory, but the cost of construction and maintenance is extremely high. Another method to bring complex ecosystems into the laboratory is much less expensive—the cost of a computer.

Models contain what researchers believe are the essential features of a system. Engineers who are designing a large and complex structure such as a dam or an airplane will often build small versions of their design first, and test these models under conditions similar to what the real structure must endure. In order for these tests to produce meaningful results, the model must incorporate all the factors that are important in its design, although on a smaller scale.

In scientific research, models also incorporate the important features of the system under study, but they are not usually miniature versions of the real thing. Instead, the model is a set of formulas or rules by which a computer simulates the real system. The formulas or rules embody the features of the system that researchers wish to study or features they must include so that the model is realistic. For example, a model of an ecosystem will probably include equations that represent at least some of the species populations—the number of individuals in each species at a given time. The populations will vary depending on environmental factors as well as interactions between species, such as predation or competition. Researchers must decide which factors are important in the model and how to construct the formulas that represent this activity. The data comes from investigations such as the nitrogen isotope study described above; without this data, modeling would not be possible.

A scientific model must include all the necessary features but must also be simplified enough to be understandable, and fast enough to run on a computer in a reasonable period of time—usually a few hours, or perhaps overnight (which is best, since few other people are using the computer at this time and it will run faster). Researchers often must write the software themselves, unless a suitable computer program already exists. If a model is too simple, it does not return meaningful results because it does not mimic the real system; if the model is too complicated, the computer will print out such a complicated mass of data that researchers will have trouble deciphering the results of the simulation. Models are only as effective as the researchers who design them.

The main benefit of computer models and simulations is that they can be used to test hypotheses and conduct "experiments" that are difficult or impossible to do in the field or the laboratory. Researchers can ask questions such as, "What will happen to this ecosystem if species X is decimated by environmental pollution?" and then run the model. The result does not prove what will happen in such a case, but will guide future research. Such guidance, whether it comes from observation or modeling, can be crucial, as in the case where Tilman and his colleagues noticed the effects of the drought and decided to further explore the issue.

VIRTUAL ECOLOGY

Computer simulations in ecology are sometimes called virtual ecology, similar to the term *virtual reality,* which describes advanced computer interfaces that simulate conditions with as much realism as possible. Virtual ecology is particularly useful in helping researchers understand what could happen as biodiversity decreases, and species extinction obviously does not lend itself to experimentation.

Real world ecosystems may be composed of hundreds and even thousands of species, forming networks of interactions and elaborate food webs. Although models incorporate information scientists have gathered over the years, some simplifying assumptions and equations must be employed. But the model can be tested with conditions that have been observed in the past, and its results compared with the real world. If a model's depiction of past events is accurate, scientists become more confident that the model offers a realistic portrayal.

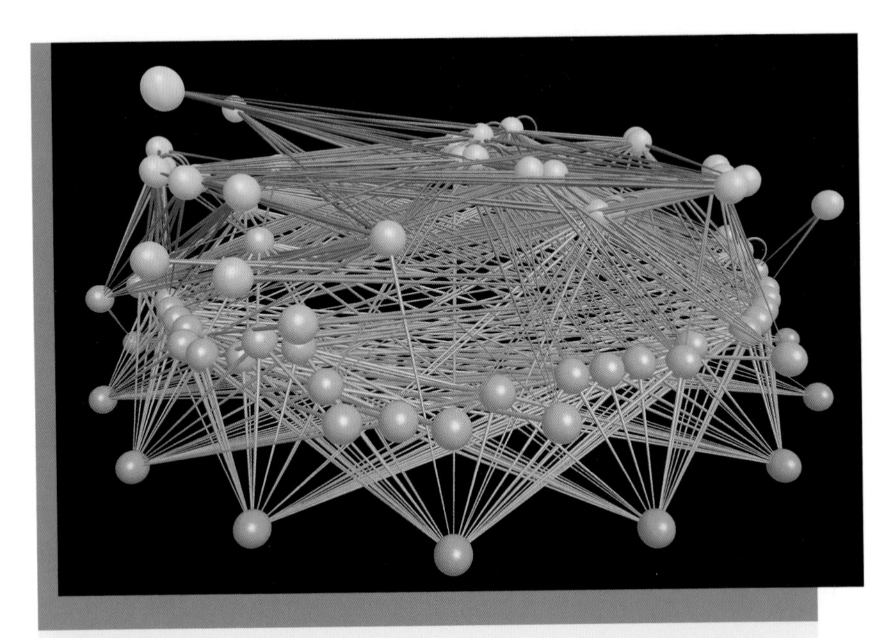

Food webs visually represent the rich variety of trophic interactions of an ecosystem. The nodes represent species, and links represent interactions between the connected species. *(Pacific Ecoinformatics and Computational Ecology Lab)*

Using food web models for 16 different ecosystems, Jennifer A. Dunne, Richard J. Williams, and Neo D. Martinez, then at San Francisco State University in California, studied what happens when one or more species is lost. In their 2002 report, "Network Structure and Biodiversity Loss in Food Webs: Robustness Increases with Connectance," published in *Ecology Letters,* the researchers showed that the effect of extinction depends not only on the number of species lost but also on their function in the ecosystem. The function in terms of the food web is represented by the organism's position and number of interactions; a web is often drawn as a network, with species represented by boxes (nodes) and lines between the boxes representing trophic interactions (prey, predator, or parasite relationships). Species with many interactions are said to be highly connected.

Dunne, Williams, and Martinez discovered that food webs suffer fewer effects from the loss of a species when the organisms are generally more highly connected, even if the food webs do not contain a wide

Pacific Ecoinformatics and Computational Ecology Lab (PEaCE)

Ecoinformatics is the use of computers and other tools to analyze information in ecology and environmental science. Computational ecology is the study of ecological systems and interactions using mathematics, statistics, and models based on data obtained from field observations and experiments. PEaCE investigates the complex networks of ecosystems, develops computer programs, analyzes data, and educates the public about ecology while promoting awareness of ecological issues. The institute partners with researchers at universities, foundations, and other research institutes, such as the Rocky Mountain Biological Laboratory in Gothic, Colorado. The current director of PEaCE is Neo Martinez.

Projects include an ambitious effort to catalog food webs and post the information on the Internet (the name of this project is Webs on the Web). Experts in ecology and computer science are developing the programs with which the data can be visualized. By widely distributing information about food webs, PEaCE fosters research and collaboration among ecosystem scientists.

PEaCE is not interested just in today's food webs. Studying environments of the past can reveal much about how ecosystems evolve and change, and this is especially important considering the drastic changes Earth is experiencing due to the encroachment of civilization and a burgeoning human population. These long lost webs are known as "paleo food webs"—*paleo* comes from a Greek word meaning long ago—and PEaCE relies on fossils to reconstruct ancient trophic interactions.

variety of species. But there is a point at which food webs begin to fall apart when highly connected species are removed, resulting in an ecosystem disaster.

These researchers currently belong to a nonprofit research institute, Pacific Ecoinformatics and Computational Ecology Lab (PEaCE), which formed in 2004 and is based in Berkeley, California. PEaCE researchers are engaged in projects that help explain and understand ecological complexity, as discussed in the sidebar on page 129.

The research of Dunne, Williams, and Martinez, along with that of other scientists, suggests that while ecosystems can adjust to changes and continue to function normally, there is a breaking point at which the ecosystem suddenly collapses. This process is similar to the familiar proverb about the straw that broke the camel's back. A single straw would seem to make little difference in a camel's cargo, but even the strongest camel has its limits. When the camel reaches its limit and can carry no more weight, an insignificant addition such as a straw will bring it to its knees.

THE ENVIRONMENT AND BIODIVERSITY

Ecological research shows that ecosystems can be robust—adaptable and adjustable—but also fragile, depending on the prevailing conditions and on the species affected. This knowledge is critical to conservation efforts. People who wish to preserve the environment against threats such as decreases in biodiversity must realize the systems they are working with are highly complex. When species have so many interactions, tinkering with one part of the ecosystem can affect all the rest of it in complicated ways.

In 2006, science writer Henry Nichols reported on the reintroduction of bison into the Lenskiye Stolby National Park, located in the Republic of Yakutia (also known as the Sakha Republic), which is a part of the Russian Federation. This area is located in northeast Siberia. The climate is frigid, and the ground is permafrost—frozen much of the year. Writing in *PLoS Biology* (the biology journal of the Public Library of Science, which maintains free public access to the scientific articles it publishes), Nichols discussed how "rewilding"—the reintroduction of formerly abundant species—may impact ecosystems.

The Yakutia rewilding involved 30 bison, 15 males and 15 females, a gift from Elk Island National Park in Alberta, Canada. Bison once roamed Yakutia in the millions, but excessive hunting thinned their population. In the mid-19th century, the completion of a transcontinental railroad across Russia gave hunters further access to the region, and the animals virtually disappeared.

Scientists believe one of the effects of this disappearance was a loss of carbon in the soil, resulting in its release into the environment, which many people think contributes to global warming. The soil also loses some of its fertility. The link between the disappearance of bison and the carbon is the grazing behavior of the bison, which break the snow cover as they forage. Snow is insulating; when it is intact, the ground retains its warmth and the ice melts, but a disruption of the snow cover exposes the ground to the cold air. Without the bison grazing, the snow remains intact and the permafrost melts, which researchers such as Sergei Zimov, director of the Northeast Science Station in Yakutia, believes could allow microorganisms to break down the carbon in the ground. The return of the bison may cause the carbon to stay where it is, preventing any possible contribution to global warming and maintaining soil fertility. (Of course, the bison add a little fertilizer of their own.) As Nichols reported in the paper, Zimov forecasts that "rewilding will increase the bioproductivity and biodiversity of the landscape."

Reintroduction of animals into ecosystems where they have vanished is an expensive and time-consuming project. An important function of environmentally conscious workers is to stop the loss of a species in the first place. This was the main goal of the Endangered Species Act, passed by the United States Congress in 1973 and signed into law by then president Richard Nixon. Today, the Fish and Wildlife Service and the National Oceanic and Atmospheric Administration Fisheries jointly manage the program.

The Endangered Species Program maintains a list of endangered species, whose numbers have dwindled enough to be in danger of extinction, and a list of threatened species, which may become endangered in the near future. Adding (or removing) species from these lists is a painstaking process, involving the collection and review of a large amount of information on the organism's population, range, history, and habitats. Government scientists identify which species may need to be listed, but any citizen may petition the secretary of the interior to add or remove a species on either the endangered or threatened list. (The secretary of the interior heads the Department of the Interior, to which the Fish and Wildlife Service belongs.)

A bald eagle in flight *(Frank Leung/iStockphoto)*

Listing has important consequences. The federal government cannot engage in any activity that imperils a listed species, and restrictions are imposed on ownership, selling, or transportation of these plants and animals. The Fish and Wildlife Service also develops a plan for the recovery of these species, and the government is authorized to make land purchases to preserve vital habitats. As of April 2009, the totals for the United States are 308 species on the threatened list and 1,009 endangered.

Despite conservation efforts, biodiversity remains under pressure. Continued habitat loss and the ever growing human population will continue to crowd Earth's plants and animals. No one is sure what effects this will have, but the researchers mentioned in this chapter, along with thousands of others, are making progress.

There is some reason for optimism. The population of the bald eagle, the symbol of the United States, had dropped to a perilously low 400 breeding pairs in the continental section of the country (consisting of all states except Alaska and Hawaii) in 1963. Thanks to a variety of programs, including the banning of a harmful pesticide known as DDT and protection offered by the Endangered Species Act, these animals have recovered to 10,000 breeding pairs by 2007. On August 8, 2007, the bald eagle officially flew off the list of endangered and threatened species.

CONCLUSION

The millions of species inhabiting Earth's biosphere are the product of many years of evolution. Members of every species are successful in finding food and generating offspring, but one species, humans, has succeeded to an overwhelming degree. The spread of cities and civilization has afforded proper habitats for human beings but only at the expense of habitats for a large number of other species, many of which are in danger of extinction or have already disappeared. This loss deprives people of enjoying the full spectrum of life on Earth. In addition, ongoing research indicates that decreases in biological diversity may have serious consequences for the productivity and stability of ecosystems, threatening essential resources such as food and fresh water.

The complexity of ecosystems makes it difficult for researchers and conservationists to know the best methods of intervention. A reintroduction of bison and their grazing habits may have beneficial effects on Yakutia's soil, as discussed above, but a single species may or may not cause a significant change, and probably will not do so until the population recovers. Some researchers are wondering if it is possible to do more, perhaps even to manage a whole habitat in the wild.

In at least one case the answers seems to be yes. Cousine Island is in the Seychelles archipelago, a cluster of 115 islands in the western Indian Ocean. Cousine is a small island of about 0.1 square miles (0.26 sq. km) in area. For decades Cousine had been home to coconut cultivators who were not careful to preserve the island's natural environment. As a result, wild pigs roamed the island, devastating the native plant species, and invasive plants dominated the landscape. Then in 1992, a South African businessman bought the island, and funded an ecological recovery.

Michael Samways, an ecologist at the University of Stellenbosch in South Africa, and his colleagues on the island began to clean up. They rounded up the pigs, cats, and chickens, and plucked the invading plants. The canopy formed by tree cover was renewed with the planting of mapou trees, which also provide ideal nesting sites for birds such as terns and noddies, whose populations began to recover. By 2006, the island hosted about 90,000 pairs of noddies. This is particularly important because of the noddy's role in the ecosystem—its droppings enrich the soil, creating an excellent environment in which native plants can thrive. Today the island is a lush tropical resort, containing four villas for tourists (who have a lot of money to spend—these exclusive villas are not cheap). By allowing a limited amount of tourism, the island's managers permit people to enjoy the island without ruining its revitalized habitats.

Conservation such as this on a global scale would be splendid, but not realistic. Scientists can never hope to have as much control over the planet as they can exert over a private, isolated island, nor would rigid control be particularly desirable. Yet the global climate changes that have been documented recently, along with mounting pollution, will continue to threaten Earth's biodiversity. The consequences are still not fully known, making ecology and research on biodiversity more important than ever.

CHRONOLOGY

1735 C.E. Swedish botanist Carl von Linné (Carolus Linnaeus) (1707–78) publishes the first edition of *Systema Naturae* (System of nature), in which he classifies species into various kingdoms, classes, orders, and genera.

1859 British scientists Charles Darwin (1809–82) and Alfred Wallace (1823–1913) propose the theory of evolution, which describes biodiversity as the result of species evolving and adapting in response to the environment and to competition from other plants and animals.

1903 In response to the growing loss of wildlife habitats in the United States, President Theodore Roosevelt establishes the Pelican Island National Wildlife Refuge, which is located on the Atlantic coast of Florida. This is the first national wildlife refuge.

1948 In the wake of an international conference at Fontainebleau, France, the International Union for the Conservation of Nature forms. This organization is now known as the World Conservation Union and is the world's largest conservation network.

1958 British biologist Charles S. Elton (1900–91) publishes a book titled *Ecology of Invasions by Animals and Plants,* which includes a discussion of the idea that simple ecosystems are less stable than diverse ones.

1973 The United States government passes the Endangered Species Act, identifying and protecting plants and animals in danger of extinction.

Australian-British scientist Sir Robert May's (1936–) book *Stability and Complexity in Model Ecosystems* summarizes his influential mathematical models that suggest higher ecosystem diversity tends to make individual species less stable.

1975 British biochemist Frederick Sanger (1918–) develops a method of sequencing DNA that will be used later in a variety of research programs, including those involved in identifying and classifying species.

1991 Biosphere 2, an ecological enclosure of 3.14 acres, is completed.

1994 University of Minnesota scientist David Tilman and Iowa State University scientist John Downing publish the article "Biodiversity and Stability in Grasslands" in *Nature,* presenting evidence that biodiversity is important in ecosystems, and initiating a heated debate among ecologists.

1996 BIODEPTH (Biodiversity and Ecological Processes in Terrestrial Herbaceous Ecosystems), a European project to test the effects of biodiversity, begins and runs for three years. The findings are similar to those of Tilman and his colleagues.

2007 Thanks to conservation efforts, the bald eagle's population in the wild is sufficient to warrant its removal from the endangered and threatened species lists maintained by the United States Fish and Wildlife Service.

FURTHER RESOURCES

Print and Internet

American Institute of Biological Sciences. "Issues in Biodiversity." Available online. URL: http://www.actionbioscience.org/biodiversity/. Accessed April 1, 2009. The American Institute of Biological Sciences, a nonprofit association committed to advancing biological research and education, provides links to a variety of interesting essays and papers on biodiversity.

Annenberg Media. "Expert Interview Transcripts: G. David Tilman." Available online. URL: http://www.learner.org/channel/courses/biology/units/biodiv/experts/tilman.html. Accessed April 1, 2009. In the biodiversity unit of Annenberg's Rediscovering Biology Web site, the interview with University of Minnesota ecologist G. David Tilman offers much insight into his ideas and research.

Balick, Michael J., and Paul Alan Cox. *Plants, People, and Culture.* New York: Scientific American Library, 1996. This well illustrated book documents the many ingenious ways people have put plants to use.

Conservation International. "Investigate Biodiversity." Available online. URL: http://investigate.conservation.org/xp/IB/. Accessed April 1, 2009. Conservation International, a nonprofit organization, provides information on the basics of biodiversity as well as related science projects and expeditions.

Convention on Biological Diversity. *Global Biodiversity Outlook 2.* March 20, 2006. Available online. URL: http://www.cbd.int/gbo2/. Accessed March 31, 2009. This report examines the present state of biodiversity and the causes of recent biodiversity losses.

Dunne, Jennifer A., Richard J. Williams, and Neo D. Martinez. "Network Structure and Biodiversity Loss in Food Webs: Robustness Increases with Connectance." *Ecology Letters* 5 (2002): 558–567. The researchers show that the effect of extinction depends not only on the number of species lost but also on their function in the ecosystem.

Fish and Wildlife Service. "The Endangered Species Program." Available online. URL: http://www.fws.gov/endangered/. Accessed April 1, 2009. The United States Fish and Wildlife Service is a bureau of the Department of the Interior that plays a major role in managing the endangered species program, as established by the 1973 Endangered Species Act. This Web resource provides a wealth of news and information about the program and the protected species.

Huston, Michael A. "Heat and Biodiversity." Letters in *Science* 299 (January 24, 2003): 512. Huston criticizes aspects of the relationship between biodiversity and ecosystem productivity.

In-cites.com. "An Interview with: Dr. David Tilman." August 2001. Available online. URL: http://www.in-cites.com/scientists/dr-david-tilman.html. Accessed March 31, 2009. Tilman, a prominent ecologist, discusses his work.

Levin, Simon A. *Fragile Dominion: Complexity and the Commons.* New York: Basic Books, 2000. The author, a theoretical ecologist, discusses how ecosystems work, why life on Earth is so diverse, and the trouble that may be caused by decreasing biodiversity.

Nicholls, Henry. "Restoring Nature's Backbone." *PLoS Biology.* June 13, 2006. Available online. URL: http://dx.doi.org/10.1371/journal.pbio.0040202. Accessed April 1, 2009. Nicholls reports on efforts to restore ecosystem biodiversity.

Novacek, Michael J., ed. *The Biodiversity Crisis: Losing What Counts.* New York: New Press, 2001. This anthology of essays, brought together by the American Museum of Natural History, includes basic information on biodiversity, the extinctions that are presently occurring, and how governments and concerned citizens can slow or stop the losses.

Patent, Dorothy Hinshaw. *Biodiversity.* New York: Clarion Books, 2003. Written for young adults, this book describes the author's experiences in Costa Rica, among other places, and explains in simple terms the interconnectedness of ecosystems and the importance of biodiversity.

Rosenzweig, Michael L. *Win-Win Ecology: How the Earth's Species Can Survive in the Midst of Human Enterprise.* New York: Oxford University Press, 2003. In this entertaining and thoughtful book, Rosenzweig relates some of the conservation movement's successes.

Tilman, D., and J. A. Downing. "Biodiversity and Stability in Grasslands." *Nature* 367 (January 27, 1994): 363–365. Tilman and Downing use statistics to find a relationship between biodiversity and stability in an ecosystem.

Tilman, David, Peter B. Reich, Johannes Knops, David Wedin, Troy Mielke, and Clarence Lehman. "Diversity and Productivity in a Long-term Grassland Experiment." *Science* 294 (October 26, 2001): 843–845. The researchers found that diverse plots performed better than even the most productive plot consisting of a single species.

Zanden, M., Jake Vander, Brian J. Shuter, Nigel Lester, and Joseph B. Rasmussen. "Patterns of Food Chain Length in Lakes: A Stable Isotope Study." *The American Naturalist* 154 (1999): 406–416. The researchers studied trophic levels in 14 lakes in Canada with the use of stable isotopes.

Web Sites

Convention on Biological Diversity. Available online. URL: http://www.cbd.int/. Accessed April 1, 2009. At a 1992 meeting known as Earth Summit in Brazil, officials from 172 governments of the world, as well as thousands of representatives from non-governmental organizations, outlined strategies for achieving sustainable development of Earth's resources. The Convention on Biological Diversity was one of the agreements to come out of this meeting, the goals of which are discussed at this Web site, along with news and developments in biodiversity.

Nature Conservancy. Available online. URL: http://www.nature.org/. Accessed April 1, 2009. The Web site of the Nature Conservancy, a conservation organization founded in 1951, describes its preservation efforts all across the globe, in every state in the United States and in more than 30 other countries.

Pacific Ecoinformatics and Computational Ecology Lab. Available online. URL: http://www.foodwebs.org/index.html. Accessed April 1, 2009. The Web site of the PEaCE laboratory contains pages on the researchers and their publications, the laboratory's current projects, and a gallery of food web diagrams.

World Conservation Union. Available online. URL: http://www.iucn.org/. Accessed April 1, 2009. The World Conservation Union is the largest conservation organization and is responsible for the Red List, a widely cited list of endangered species. Its Web site provides news, information, policy documents, and conservation publications.

5

The Biology and Evolution of Viruses

To soldiers who survived the ghastly combat of World War I and its machine guns, toxic gases, artillery shells, and deadly mines, succumbing to a tiny organism invisible even to optical microscopes might have been unthinkable. Yet beginning in 1918, as World War I was concluding its four horrible years, many combatants, veterans, and millions of other people died in a *pandemic*—a worldwide epidemic that was one of the worst in history. The initial symptoms were hardly worse than a cold. But death followed quickly, and by 1919, when the disease finally subsided, more than 50 million people were dead, including about 600,000 Americans—10 times the number of American soldiers who perished in combat on the European battlefields of World War I. Almost 200,000 Americans died in the month of October 1918 alone. John M. Barry, author of *The Great Influenza,* notes that the pandemic "killed more people in a year than the Black Death of the Middle Ages killed in a century."

This epidemic is sometimes known as the Spanish flu because it claimed many lives in Spain and received a great deal of publicity in that country. But the earliest outbreak apparently occurred in the vicinity of Fort Riley, Kansas, in March 1918. Because the war dominated the news headlines at this time, and the number of deaths was not yet large, this and other early outbreaks escaped much attention.

Influenza, also known as the flu, is due to a *virus.* The influenza virus infects the respiratory tract (nose, throat, and lungs) and causes fever, headache, coughing, a sore throat, stuffiness, and body aches. Each

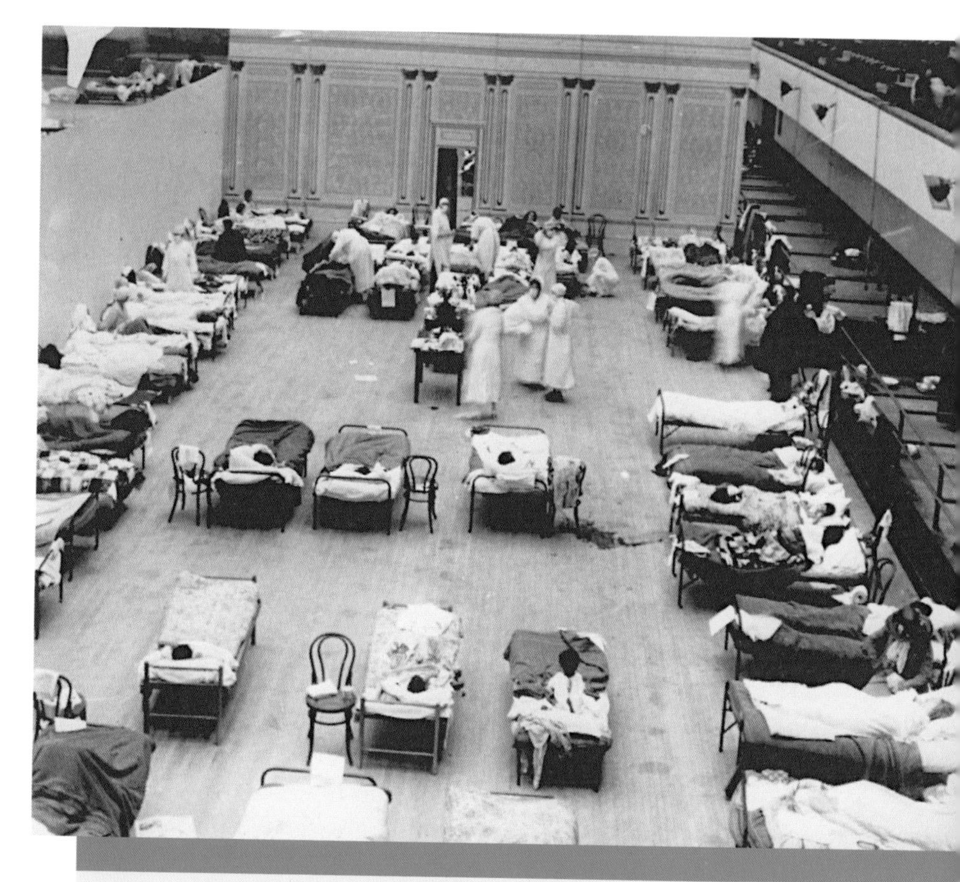

American Red Cross volunteers tend to influenza victims in Oakland, California, in 1918. *(Edward A. "Doc" Rogers/Joseph R. Knowland Collection at the Oakland History Room, Oakland Public Library)*

year millions of people contract influenza, and in the United States the primary "flu season," when a lot of cases arise, lasts from December through March. (No one is certain exactly why this is the peak season for the flu.) Although influenza can be fatal to the elderly and the very young—and each year there are thousands of fatalities—most flu seasons pass without a severe outbreak. But every so often the influenza virus mutates into an especially dangerous organism, as it did in 1918.

Other viral diseases have also appeared from time to time. Dangerous ones include acquired immune deficiency syndrome (AIDS), severe acute respiratory syndrome (SARS), Ebola hemorrhagic fever, and others that have appeared recently, along with older ones such as polio and smallpox. The viruses responsible for these diseases are but a few of a group of widely diverse infectious agents, only a small number of which pose any threat to humans. What makes certain viruses so deadly depends on the virus and how it evolves. This chapter describes the evolution of viruses and how the human body has evolved to fight back, resulting in a constant struggle as viruses and humans strive to outfox the other. Viral evolution is an active area of research—a frontier of science that delves into the nearly invisible world of viruses, some of which have the capacity to kill on a scale comparable to nuclear weapons.

INTRODUCTION

The germ theory of disease was not widely accepted until the 19th century. French chemist Louis Pasteur (1822–95) proved in the 1860s that microorganisms were present in air, and in the 1870s and 1880s, German biologist Robert Koch (1843–1910) and others discovered the specific germs that cause diseases such as anthrax, tuberculosis, and cholera. These germs were bacteria, small single-celled organisms that seek warm, moist, food-rich environments in which they can grow and proliferate. Plants, animals, and the human body offer just such an environment.

In the hope of isolating other disease-causing bacteria, researchers often strained infected tissues with filters, trapping the microorganism. In 1892, Russian researcher Dimitri Ivanowsky (1864–1920) filtered the sap from tobacco plants afflicted with tobacco mosaic disease, which inhibits growth and causes spots on the leaves, making them appear decorated like a mosaic. But the agent that caused tobacco mosaic disease was too small to be trapped by the filter, and Ivanowsky did not know what to make of this result. Dutch botanist Martinus Beijerinck (1851–1931) repeated the experiments a few years later and showed in 1898 that the agent could reproduce—its ability to infect did not diminish over time, which would be expected unless the agent had some means of reproduction. Beijerinck called it a virus, a Latin word meaning poison.

Bacteria tend to be small cells, and many are about 10–50 times smaller than the average size of a human cell. Viruses are about 10 times smaller than bacteria, and range from 0.0000008 to 0.000012 inches (20–300 nm). Most particles in this size range are far too small to be seen even with the best optical microscopes, since the optical phenomenon known as diffraction—the bending of light rays traveling through a narrow opening—limits the ability of light-based microscopes to resolve small objects.

But electron microscopes offer superior resolving power. These microscopes use a beam of tiny negatively charged particles called electrons, which are focused by electromagnetism. Although electron microscopes also have their limits, they can produce magnification of up to a few million times (compared to a maximum of about 2,000 for light-based microscopes). German researchers Ernst Ruska (1906–88) and Max Knoll (1897–1969) built the first electron microscope in 1931, and in the 1930s biologists began to create images of the tobacco mosaic virus. Since then, virologists—scientists who study viruses—have used electron microscopes as well as other techniques to visualize these minuscule objects.

There is a great deal of difference between bacteria and viruses besides just size. A bacteria is a cell, containing the nutrients and molecules that make it capable of independent life. Viruses infect cells—either bacteria or the cells of plants or animals—and cannot survive for long outside of these host cells. A virus is an obligate intracellular parasite—it survives and reproduces only by invading the interior of a cell and using that cell's molecular machinery to make more viruses. Viruses consist of only a few genes encapsulated in protein and, in some viruses, a membrane.

As described in chapter 2, genes contain instructions to make specific proteins. In cells, these instructions are in the gene's sequence of double stranded DNA. The enzymes and other molecules involved in reading genes and making proteins are taken over by the invading virus, even though the genes of some viruses are made of RNA instead of DNA. The genes of the virus do not code for anything but a few proteins that the virus needs for the coat, to invade the cell, and to take over or complement the host cell's enzymes. A successful infection means that the virus makes many more copies of its genome—its set of genes—and the proteins of the coat, which assemble into new virus particles that exit the cell and attempt to find their own host. (Computer viruses—programs that infect computers and usually replicate to infect other

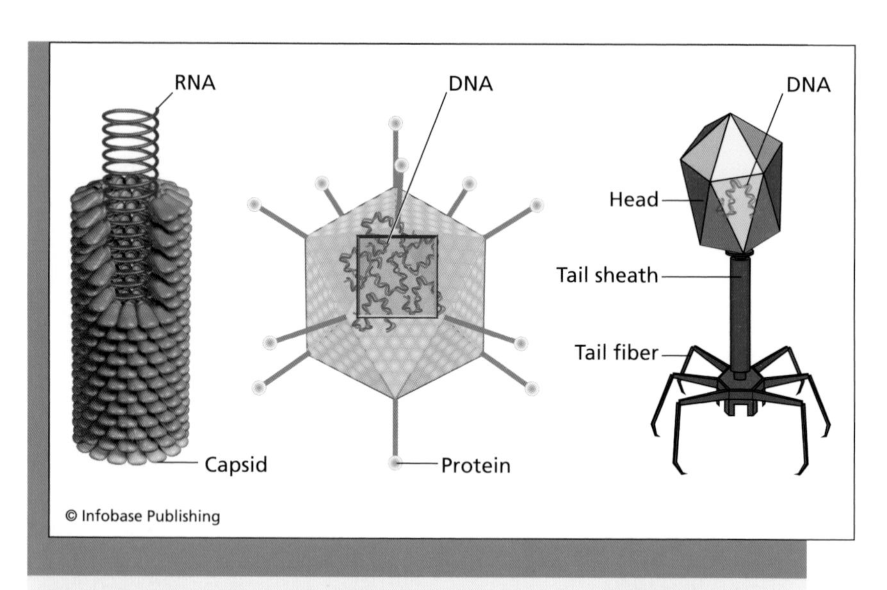

Some viruses, such as the tobacco mosaic virus, have a cylindrical shape, as shown on the far left. Other viruses, such as a group known as adenoviruses, have the shape of a polyhedron, as shown in the center. Bacteriophages—viruses that infect bacteria—such as the virus on the far right, have a shape similar to a space probe; the virus "lands" on a bacterium and injects its DNA through the cell wall.

computers—were given their name because of the similarities between these programs and biological viruses.)

The protein coat, known as a *capsid,* is important to protect the virus's genes. Unprotected RNA and DNA can suffer damage that renders them unusable, so viruses shield their nucleic acid with a capsid that can be as strong as hard plastic, but is also flexible. The capsid can be made out of dozens of copies of a single protein or it can be made out of multiple proteins, but either way it tends to have one of a small number of shapes. Examples of the common shapes are shown in the figure above.

Since a virus relies on another object for most of the resources needed for life, including reproduction, is a virus alive? Biologists do not always agree on the answer. Perhaps a virus can be considered potentially alive, capable of coming to life when it infects a cell. As Alan J. Cann, author of *Principles of Molecular Virology,* notes in his textbook, "One view is that inside the host cell viruses are alive, whereas outside it they are merely complex assemblages of metabolically inert chemicals."

If a virus is successful, it proliferates after invading the host. Sometimes this causes little harm, but in other cases, such as in AIDS, polio, and smallpox, the consequences are dire.

But hosts are not defenseless. To get inside the cell, a virus usually needs to latch onto a molecule on the cell's surface; this molecule, called a receptor for the virus, is often a protein, and is the means by which a virus recognizes its victim. (These receptors perform a variety of useful functions for the cell, otherwise they would not be present—viruses are merely taking advantage of their presence.) But the protein that the virus needs to recognize its receptor can also be one of the means by which a host recognizes its viral enemy.

Argonne National Laboratory researchers determined the structure of this virus's coat, which looks somewhat like chain mail. *(Argonne National Laboratory Media Center)*

Antibodies

Vertebrates (animals with backbones) have a number of defense mechanisms against invaders such as bacteria and viruses, but the most complex defense is the immune system. This system is spread out over the whole body, and consists of millions of cells called lymphocytes, which are a kind of white blood cell. Humans have about two trillion lymphocytes circulating in the blood and lymphatic system (a network of vessels that carry extracellular fluid), as well as in various organs and tissues. The immune system's job is to distinguish between molecules that belong in the body and those that represent an intruder. To do this, the immune system learns early on to recognize certain molecules on all of the body's cells. Anything new is considered an enemy.

The immune system "sees" molecules by their shape. Proteins fold into a conformation based on the amino acid sequence, and other large molecules also have a shape. (The immune system does not generally recognize small molecules.) In the course of development, the human body generates a few million slightly different types of lymphocytes, each of which are genetically programmed to recognize a

Recognizing the virus is important because it allows the host to mount a defense. For example, in humans the immune system contains cells that distinguish viral invaders, and mark these viruses for destruction. A viral invasion initiates a war between the virus particles, which are reproducing as fast as they can, and the immune system, which is destroying them. But the immune system is at a disadvantage because it takes time for it to call up the molecules and cells that can recognize and kill the virus—unless the immune system has encountered the virus before. The immune system has a memory and it does not forget a "face"—the virus's exterior molecules—that it has had to fight specific molecular shape. When a lymphocyte encounters a molecule it is programmed to recognize, it launches an immune response. Lymphocytes known as B lymphocytes (or B cells) produce proteins called antibodies that latch onto the invader and mark it for destruction by other cells. Any substance that elicits an antibody response in the body is known as an antigen, which is short for antibody generator. Antibodies bind to a specific antigen and they do so in a specific way. In the case of a viral invader, antibodies usually recognize a specific protein protruding from the virus's surface.

When an invader stimulates specific lymphocytes to produce antibodies, other lymphocytes are generated that remember this antigen. Although these cells are not essential to fight off the initial infection, the memory is important because it identifies invaders that the body has had to fight before, and thus may need to fight again. The existence of these "memory" cells increases the speed and effectiveness of the antibody reaction in the next encounter with this invader. But if the invader changes its shape, the circulating antibodies may no longer recognize it. When this happens, the memory is not invoked and the immune system must fight the invasion from scratch.

before. This defense involves proteins known as *antibodies,* which are described in the above sidebar.

Physicians realized long ago that an introduction of a mild dose of the virus can help a person to fight off a later, more serious infection from that virus. This technique was called variolation. (The physicians did not know how it helped the immune system, but knew only that it tended to work.) Variolation goes all the way back to China in 1000 B.C.E.; some areas in China suffered from epidemics of smallpox, a disease with a high mortality, and the Chinese would sometimes expose themselves to a small amount of infected tissue, such as the dried crusts of sores from persons infected with the disease. (Smallpox is also known as variola, which is where variolation gets its name.) The technique eventually spread to Europe and the Americas, but since the mild dose could easily progress to a severe case, variolation was risky. Benjamin Franklin (1706–90), for example, decided not to administer variolation to his son Francis because of the risk—but the child contracted smallpox and died in 1736. In his autobiography, Franklin wrote, "I long regretted bitterly, and still regret that I had not given it to him by inoculation." Franklin later became an advocate for variolation, as were Thomas Jefferson (1743–1826) and George Washington (1732–99).

Vaccination, the modern process of generating immunity, is much less risky, and exposes a person to viruses or viral molecules that have been rendered harmless. This technique began in 1796, when British physician Edward Jenner (1749–1823) used material from cowpox sores to vaccinate people against smallpox. Cowpox usually affects cows but is sometimes contracted by farmers, in which case the disease is usually quite mild. Yet people who have had cowpox rarely get smallpox, prompting Jenner to try his vaccine experiment. (The term *vaccine,* although now applied generally, was named specifically for Jenner's technique—the term derives from *vacca,* a Latin word for cow.)

Although vaccination can be effective, physicians must find a safe dose of an appropriate material to establish an immune system memory. This is not always simple, and depends on the nature of the virus.

CLASSIFICATION OF VIRUSES

There are several ways to characterize viruses, but one of the most common is to classify them according to their genetic material. For viruses, genes can be coded in either DNA or RNA. These molecules are nucleic acids, composed of covalently bonded units known as nucleotides or bases connected together in a chain or strand. Because of weak bonds known as hydrogen bonds, two strands can wind around each other, as in the double-stranded DNA helix (see the figure on page 42). The genetic material in viruses can be single-stranded RNA, single-stranded DNA, double-stranded RNA, or double-stranded DNA. Examples of viruses in each category have been found, but double-stranded DNA and single-stranded RNA are the most common. For instance, the virus that causes smallpox contains double-stranded DNA, as does a family of viruses known as papovavirus, which among other things cause warts. Single-stranded RNA viruses include human immunodeficiency virus (HIV), the virus that causes AIDS, as well as the influenza viruses, poliovirus, and rhinovirus (which gives people the common cold).

Since viruses rely on the cells they infect for most of their needs, they carry only the genes that are essential for their reproduction but are not found in the host cell. This means viruses require few genes—the genome with the smallest number of genes is the hepatitis B virus (which infects the liver), which has only four genes! The largest, Mimivirus, has 900. Compare that with the human genome, which has roughly 20,000 genes.

The primary goal of all viruses, regardless of their type, is to find a hospitable cell and make as many copies of itself as possible. Genes of DNA viruses are already in the right form—the genetic material in cells is made of DNA, the sequence of which contains the instructions by which enzymes produce proteins. The enzymes first transcribe the gene into RNA, then other enzymes translate the RNA sequence and produce a protein with the proper sequence of amino acids. When a cell reproduces, it divides in two, prior to which it replicates its DNA so that both daughter cells will have a copy. DNA viruses slip their genes into these transcription and replication processes, as shown in the figure on page 150, pirating the cell's machinery to make the viral proteins and to replicate the viral DNA. Fortunately for viruses, the genetic code is the same for virtually all organisms on Earth, so the genes in the virus can be read by any cell.

For RNA viruses, the route to reproduction can be a little more complicated. The genes of some viruses slip into the cell's machinery as RNA templates, which are then translated into the proteins the virus needs. To make copies of their genes, viruses need enzymes to replicate RNA. Since the right enzymes are not normally found in cells, the gene or genes coding these enzymes must be part of the virus's genome.

Other RNA viruses engage in even more complicated operations. Some of these viruses contain genes coding for protein enzymes that reverse the normal flow of genetic information—instead of the usual process of transcribing a DNA sequence into RNA, these enzymes make DNA copies of the virus's RNA genes. Such a virus is known as a retrovirus (*retro* means going backward). The DNA copies can then sometimes get inserted into the cell's DNA, most of which is housed in the

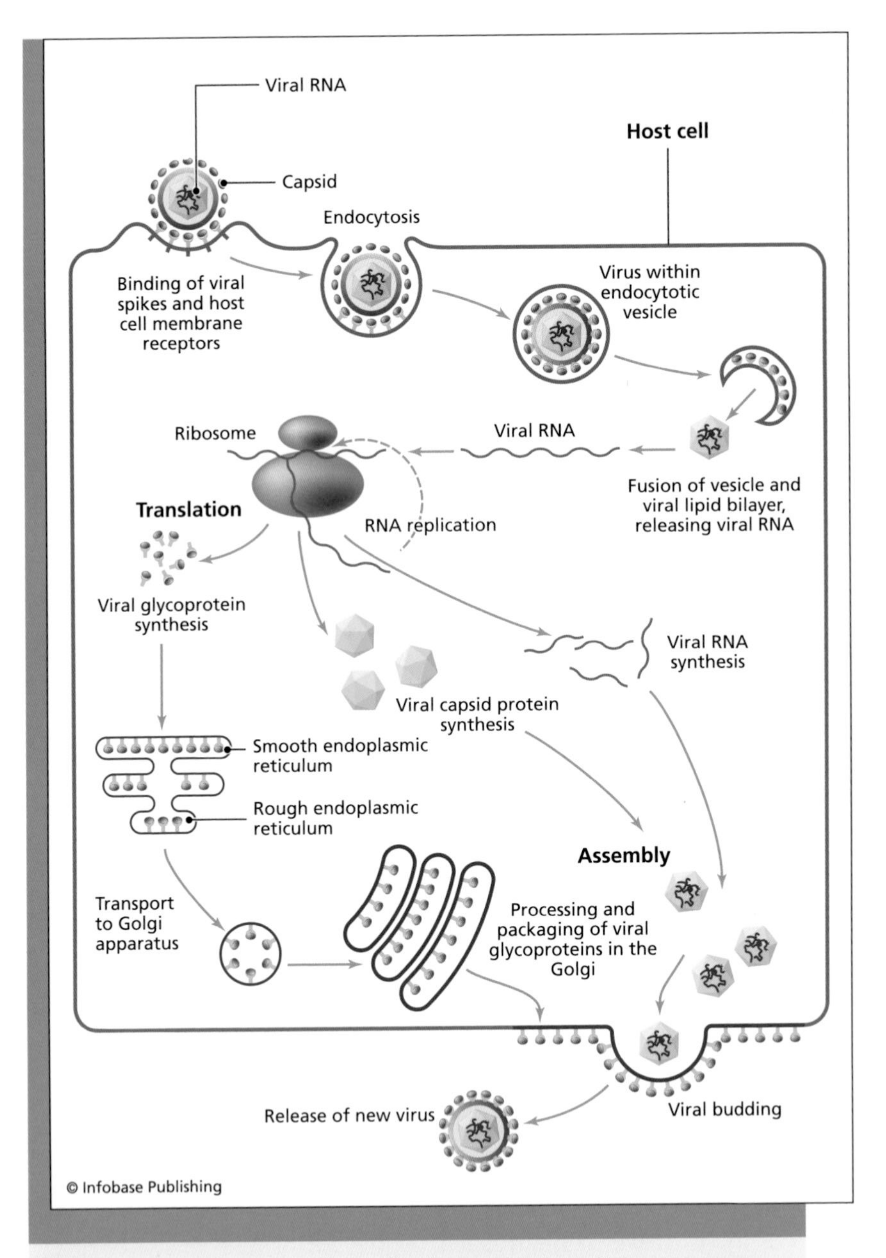

Viruses hijack the host cell's replication and transcription enzymes, using them to make more copies of the viral RNA (or DNA) and to make proteins, such as capsid proteins, needed by the new virus particles.

cell's nucleus. This is a disastrous situation for the cell, for the foreign DNA is now a permanent part of its genetic material. HIV is an example of a retrovirus and is an exceptionally difficult virus to deal with, as discussed in more detail below.

Another way to describe the nature of a virus is to consider its hosts. Most viruses cannot infect just any cell they happen to encounter, since viruses usually sneak into cells by binding to a receptor. If a cell does not have this receptor sticking out from its membrane, the virus fails to latch on, and proceeds to flow through the solutions and tissues of its host's body until it finds the right cell, or until it gets caught and eliminated.

Because of the need for a specific receptor to enter the cell, many viruses can infect only one or a few types of cells. For example, there are viruses known as *bacteriophages* that only attack bacteria (the term *phage* comes from a Greek term, *phagein,* meaning to eat). Other viruses target certain types of cells or tissues; this is the case with HIV, which infects specific types of white blood cells and brain cells.

Receptors are often proteins, and in many cases are made and expressed in cells of only one species, which means that only one type of organism contains cells with this particular protein. If this is the case, then the virus can infect only this species. This is a critical point. In such a situation, the virus is limited to a single host, which it must find and infect in order to reproduce. Without the host, the virus is nothing but a little bit of nucleic acid wrapped in a protein coat, which will eventually degrade and fall apart.

If, for example, a virus infects only humans, then the prevention of any human infection means that the virus will eventually die off and become extinct. This is what happened to smallpox. After a series of widespread vaccination programs and a concerted effort on the part of many agencies and health organizations, the World Health Assembly announced in 1980 that smallpox was extinct in the wild. Without a host to infect, the smallpox virus disappeared. Although samples of smallpox virus remain available in the freezers of scientific laboratories, the last infection to occur from natural causes happened in 1977 in Somalia.

The keys to smallpox eradication were an effective vaccination and the inability of the virus to infect other hosts. Other viruses are not so limited. Influenza, which caused the scourge of 1918 as well as other epidemics over the years, has the ability to "play the field."

INFLUENZA—BIRDS, PIGS, AND HUMANS

Chapter 2 of this volume discussed genetic variability. Although members of the same species have the same genes, there are slight variations within the species. In humans, the gene that causes eyes to have a certain color is not the same in all people; the author of this book, for example, has blue eyes, while many other people have brown or green eyes.

Genetic variation is important for change and evolution. Mutations, or changes in genetic information, can increase the organism's odds of survival and reproduction, and the new genes will get passed on to the offspring and will tend to be found in more and more of the organisms. (And if the mutation decreases an organism's odds, it will tend to die out.) Members of the same type of virus can also vary, and these variations are important for the same reasons—evolution and change.

Three types of influenza viruses are known, and have simple (and nondescriptive) names: A, B, and C. Types B and C are not generally much of a threat to humans. Type A is the killer, and is broken into subtypes on the basis of two critical proteins on the virus's surface—hemagglutinin and neuraminidase. Hemagglutinin latches onto the receptor so that the virus can slip inside the cell; neuraminidase is important for the reverse procedure, helping the newly created viruses escape from the cell. In the classification scheme, H1N2, for example, has a hemagglutinin of type "1" and a neuraminidase of type "2" (these are also not very descriptive names but simply refer to slightly different versions of the protein, which can be determined by various laboratory tests). There are 16 known hemagglutinin types and nine neuraminidase types.

One of the reasons influenza is such a problem is that type A viruses can infect not only people but also birds and pigs, among other animals. In fact, wild birds seem to be the natural hosts for these viruses—all subtypes of influenza A can infect birds, whereas only a few subtypes are currently infecting people. Wiping out influenza type A would not be possible in the same way as it was with smallpox. To eradicate influenza, all the birds and many other animals would have to be vaccinated or killed.

But the difficulties in fighting influenza run even deeper. Although there is an effective vaccination for influenza, it does not provide long-term protection. Each year, in October or November, many people get their annual "flu shots"—injections of influenza vaccination—or the nasal-spray vaccine. The reason people need to get shots each year is that the virus changes rapidly. Vaccines offer protection against infection of the types of viruses that have been found circulating among people that year. The vaccines stimulate the immune system memory and the production of antibodies, so the immune system will be primed to mount a successful fight if the person is exposed to these viruses. But next year's batch of viruses will have changed. Viruses evolve, as does all life, or potential life, on Earth.

Evolution is normally a slow process. Millions of years were required for humans to attain their present form, and the same is true for most creatures currently roaming the planet. But unlike the complex physiological systems and the huge number of interacting cells that make up a human, viruses are extremely simple. A small change in a person's genes can affect a lot of different systems because many of the parts work together and influence one another, but the activities of a virus are a lot less complicated. And because viruses have a high rate of reproduction and are so small, large numbers can inhabit any given host, which means that there is a greater chance for changes to occur. Even if a certain mutation is highly unlikely, one is bound to occur quickly if there are a huge number of opportunities for it to take place.

Different strains of influenza often crop up. A flu strain is a variety of influenza, and each subtype, such as H1N1, can have a number of strains at any given time. The genome of influenza type A viruses codes for a total of 11 proteins (two of which are hemagglutinin and neuraminidase), and mutations can affect their properties, leading to a different strain.

A major change in hemagglutinin or neuraminidase would result in a new subtype. One of the reasons why these genes have such a prominent role in classifying viruses is because their proteins are found on the virus's surface, and interact with the molecules scientists use to study viruses. But a minor change might not be noticeable—except to a certain number of antibodies. As described in the previous sidebar, antibodies are the proteins that bind to the foreign molecules in the body, marking them "To Be Destroyed." Since hemagglutinin and neuraminidase are exposed on the virus's surface, they make excellent targets for antibodies (as well as for other molecules used by scientists to latch onto the viruses). But a small change in a protein's sequence of amino acids can result in a change in its shape, and this means an antibody that once bound the protein may no longer do so. For example, a certain antibody may have clung to a particular bend that is not present anymore. If this happens, the immune system does not recognize the virus as quickly, and the defense is slower. Each year scientists formulate a new flu shot based on the strains that have been found or predicted for that year.

With the latest genome sequencing technologies, sequencing a small genome such as a virus is not difficult, once researchers isolate it. Influenza A has about 13,500 bases in its genome, which is 0.00045 percent the size of the human genome. But it is the variety, not the size of the genome, which keeps scientists busy. To provide important information to all researchers working on the influenza virus, the National Institute of Allergy and Infectious Diseases (NIAID) launched an ambitious project in November 2004 called the Influenza Genome Sequencing Project. (NIAID is a branch institute of the National Institutes of Health.) The goal is to sequence as many strains as possible. This is no small task; for instance, St. Jude Children's Research Hospital in Memphis, Tennessee, has a collection of 12,000 influenza viruses, isolated from a variety of sources. Project scientists have sequenced more than 3,000 virus isolates.

VIRAL EVOLUTION

Viruses must evolve more quickly than their hosts or the virus will become extinct. A mutation may cause a change in a cell's receptor, and if there are no viruses that can dock onto the new molecule, they will all expire. Smallpox failed to evolve fast enough after everyone was vaccinated, so once all their hosts were gone, so was smallpox, at least in the wild.

The smallpox virus is a DNA virus. RNA viruses such as influenza and HIV have an advantage over DNA viruses in terms of evolution speed—RNA viruses can evolve more quickly. One of the reasons for the advantage is that the replicating enzymes of the RNA viruses are not as accurate as those for DNA replication. As a result, mistakes occur more frequently. (This is one of the few instances where making more mistakes is better than making fewer!) The error rate for HIV, for example, can be as high as one base in a thousand. Although this does not sound too bad, it is about a million times higher than error rates in most cells, which usually make a mistake in about one base in a billion. (Since many DNA viruses use these same enzymes, the error rate for these viruses is about the same as it is for the cells.) With an error rate of one in a thousand and a genome numbering about 10,000 bases, HIV would make about 10 mistakes every time it replicates! The rates can vary, though, and some strains may have much lower rates.

With so many mutations, there is a lot of material for evolution to act upon, and the rates for influenza and HIV are comparable. In 1993, John Drake, a researcher at the National Institute of Environmental Health Sciences in North Carolina, published a paper, "Rates of Spontaneous Mutation among RNA Viruses," in the *Proceedings of the National Academy of Sciences.* For influenza type A viruses he reported a mutation rate of about one base per genome replication—one change each time the virus makes a copy of itself.

Some of these changes can have big effects. As mentioned earlier, some of these mutations will make the immune system's job a lot more difficult—the virus is elusive or even "invisible" in its new "disguise." The result is an increase in the number of people who contract influenza, at least until immune systems catch on (with the help of scientists who monitor and isolate the new viruses in order to develop new vaccines).

Mutations can also increase the mortality rate—the number of patients who die from the disease. People do not all respond in the same way to an infection because people are not all alike. Individual differences are due to differences in diet, environment, and genetics, and cause some people to get sicker than others. Some people recover, and some do not. A mild strain produces low mortality, for only the most susceptible die. Strains that are virulent—a word denoting severe harm, having the same root as virus—can overpower even the healthiest and strongest victim.

The high mutation rate of influenza viruses means that new strains appear frequently, some of which evade detection and some of which create new symptoms in the host. But influenza has yet another trick to play, and this trick makes influenza even more dangerous.

Most viruses have all their genes on a single piece of RNA or DNA. Influenza, however, has a segmented genome. The 11 genes of an influenza virus reside on eight different segments. When the virus infects a cell, the segments are copied and reassembled into a new virus. But what if there is more than one kind of virus in that cell? The segments can get scrambled, and some of the new virus particles may have a new combination of segments, some from each of the different viruses. Virologists call this process *reassortment.*

Reassortment can cause a major shift in influenza's properties, just as chromosomal shuffling is an important source of genetic change in animals. For instance, in 1956 a new subtype, H2N2, appeared in Asia. It was a new combination, a reassortment of strains, and human immune systems had never encountered it before. No one had any immunity, resulting in a pandemic that spread to the United States and elsewhere by 1957. Even worse, the strain turned out to be a killer. This pandemic, known as the Asian flu, claimed several million victims worldwide, including 70,000 Americans.

The threat of reassortment is particularly dangerous because influenza infects so many hosts. This is true even though most strains are specific to one host. As described earlier, viruses tend to be specific for one type of cell or one specific host because the target (unwittingly) expresses the receptor that the virus uses to gain entry. But different viruses may infect the same host and find themselves inside the same cell, giving reassortment an opportunity to occur.

An influenza virus subtype can also jump to a different host if a new strain emerges. This process has alarmed virologists and physicians recently because of a particular virus with type 5 hemagglutinin—an H5 virus—that normally infects only birds. In 1997, a strain of H5N1 made the leap to humans.

Wild birds may carry H5N1 but do not often get sick, though as carriers they spread the virus. Domestic poultry such as chickens also get infected, and they are not as lucky—chickens come down with an avian (bird) version of influenza and suffer a high mortality from H5N1.

In Hong Kong, thousands of chickens began to die from influenza in 1997. Then a boy was admitted to the hospital with flu-like symptoms. Yet the virus obtained from the young patient did not match any known influenza viruses—at least not those that usually infect humans. Virologists scrambled to "type" the strain—to identify the virus—and soon they realized they had a highly unusual case. The child, who succumbed to the illness, had come down with avian flu caused by an H5N1 infection.

Other cases, 18 in all, began to appear in Hong Kong. But an epidemic had not yet emerged because H5N1 was not easily transmitted from person to person. In Hong Kong, consumers often buy live poultry at markets, and it was in the course of these activities that officials suspected patients had become infected with H5N1. In other words, people were getting the disease from chickens rather than from each other. The solution to the problem was drastic: Hong Kong's authorities destroyed all the poultry in the region, well over a million birds. Killing the chickens eliminated the infection. The solution worked, and for a while H5N1 was held at bay.

Health officials all over the world applauded Hong Kong's efforts, drastic though they were. What officials feared most was that, if the infections were allowed to continue, mutations would bring about new strains, some of which might be capable of easily traveling from person to person. Since humans have little immunity to H5N1, the virus would spread rapidly, and if the strain was a killer then the resulting pandemic would be catastrophic.

But H5N1 infections have continued to strike bird populations in Asian countries and have spread to Europe. A small number of human cases have also occurred, and in 2004 a few of these in Vietnam and Thailand were apparently due to contact with an infected person, which means the virus was transmitted from person to person. But the transmission was not airborne, for it occurred only between people who were in close contact. Although cases of human transmission are alarming, the most dangerous situation occurs when viruses can travel in the air and stay attached to surfaces. This is how the strains that commonly infect humans spread so quickly. An infected person sneezes, coughs, or touches a doorknob, eventually introducing the virus to its next victim.

The biggest worry is that H5N1 could infect someone at the same time as one of the common human flu strains. If reassortment occurs between H5N1 and a common human influenza virus, the result might be a deadly new virus capable of airborne transmission. Because of this threat, health agencies in the United States such as the Centers for Disease Control and Prevention (CDC) are monitoring the spread of H5N1. CDC helps global health agencies such as the World Health Organization to identify and track this virus in both birds and humans. In CDC's assessment, posted on its Web site, the virus "continues to pose an important public health threat." In the event that a dangerous new strain emerges, CDC has prepared plans and conducted training sessions to help local, state, and federal government officials to handle the

Centers for Disease Control and Prevention (CDC)

Malaria, a serious disease caused by a microorganism carried in certain mosquitoes, used to occur frequently in the United States, especially in the southeastern portion of the country. Thousands of soldiers from both sides in the American Civil War, which was fought mostly in the southeastern states, required hospitalization because of malaria. In the early 1940s, as the United States trained its soldiers to fight in World War II, concerned officials decided to spray insecticides around the camps to kill the mosquitoes. They organized an agency called Malaria Control in War Areas to perform this task. This agency was so successful that malaria is no longer a significant threat in the United States. Realizing that this organization could fight other diseases, the government transformed it into the Communicable Disease Center on July 1, 1946, based in Atlanta, Georgia. (The southeastern location for the center was primarily because

emergency. CDC scientists and physicians have a long history of dealing with new infectious diseases, and the sidebar above provides some additional information about this important government agency.

An additional problem that arises with H5N1 and the "bird" flu has complicated attempts to create a vaccine. Many different influenza vaccines have been made in the past using well established techniques, although the vaccines must be constantly updated because influenza viruses evolve so rapidly. But the process of manufacturing flu vaccines involves growing the viruses in hen eggs. This works well for human influenza viruses, but H5N1 kills chickens, so this procedure cannot be used for H5N1.

One solution to this problem is to develop alternative methods of growing the viruses. Techniques do exist for growing viruses in small of its earlier malaria campaign.) Later, the agency became known as the Centers for Disease Control and Prevention.

Today CDC is concerned with a large number of health issues. As a component of the Department of Health and Human Services, CDC operates centers devoted to environmental health, injury prevention, global health issues, infectious diseases, public health information, terrorism preparedness, and workplace health and safety. CDC also compiles important health statistics every week and issues a report, *Morbidity and Mortality Weekly Report*. These statistics help officials to gauge the state of health in the country and to identify trends that need investigation.

Virus research is a major focus of CDC. Researchers at the Emerging Infectious Diseases Laboratory track and study H5N1 as well as many other disease-causing viruses and agents. CDC also publishes a monthly journal, *Emerging Infectious Diseases*, for researchers around the world to report their findings. The research and diligent observations conducted by CDC scientists and administrators reduce (but cannot eliminate) the risk of a serious epidemic developing in this country.

cells or bacteria, and although they may be more difficult and expensive, these methods might have an advantage in allowing faster and easier production. This advantage would be extremely important in case of a sudden H5N1 outbreak, when large amounts of vaccine would be needed in a hurry.

Another solution to the vaccine problem employs genetic techniques that scientists have developed for studying DNA. By changing the genetic composition of the virus, scientists can create strains of H5N1 that will grow in hen eggs yet produce viable vaccines. In April 2007, the Food and Drug Administration, the agency responsible for food and drug safety in the United States, approved the first H5N1 vaccine in this country. The French company, Sanofi Pasteur, which makes this vaccine, uses the egg technique. Although the company has no plans of

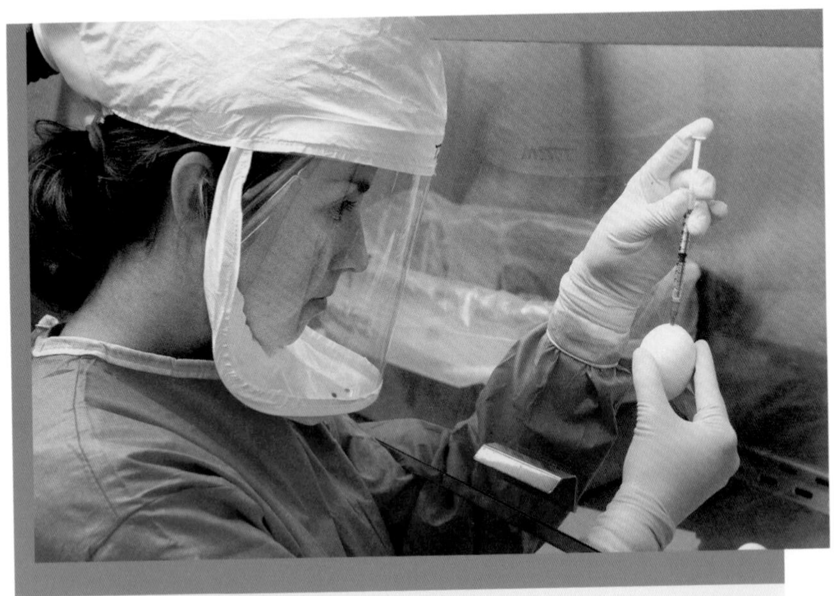

Taronna Maines, a CDC biologist, observes proper precautions as she works on H5N1. *(Greg Knobloch/CDC)*

selling the vaccine to individuals, the United States government purchased doses and added them to its stockpile in case of emergency.

AIDS EPIDEMIC

Having a vaccine that has already been tested is an important element in the fight against a widespread outbreak. Yet viruses and their ability to cause epidemics should never be underestimated. AIDS is a perfect example.

In 1981, American doctors noticed a significant increase in rare diseases among young male homosexuals in certain areas of the country. The doctors reported their findings in CDC's *Morbidity and Mortality Weekly Report*. After this important alert, physicians began finding other unusual patterns. The common factor was an impaired immune system that failed to fight off infections and other diseases. Since the syndrome—a cluster of signs that characterize an illness—was acquired later in life rather than appearing at birth, scientists called it acquired immunodeficiency syndrome, or AIDS for short.

The focus was initially on homosexual behavior as a cause of the disease, but drug abusers who shared hypodermic syringes (needles) were also affected, as were people who had recently received blood transfusions. In 1983, American researcher Robert Gallo and French researcher Luc Montaigner discovered a virus that eventually became known as Human Immunodeficiency Virus after Gallo and colleagues showed that the virus causes AIDS. This virus is transmitted by sexual contact with an infected person or by contamination with the blood of an infected person. A number of people in the 1980s, such as writer Isaac Asimov, unknowingly received infected blood during treatment and later developed AIDS, but blood donations are now screened for HIV.

The World Health Organization estimates that about 25 million people in the world have died of AIDS, and close to 40 million people are now living with the disease. Effective treatments consist of various antiviral medications that target enzymes of the virus and reduce replication. These treatments, though they are often able to hold the disease in check, are unable to rid the body of all the viruses. The problem is that HIV makes a DNA copy of its RNA and inserts this DNA into the host cell's genome. (The enzyme it uses to make the DNA copy is known as reverse transcriptase, which is generally the target of AIDS drugs.) Since the cell is unable to extract the foreign DNA, it stays where it is. Sometimes this alone can cause problems. In other cases, most of the symptoms of AIDS do not begin until after the virus has begun to replicate in earnest, which may take years.

HIV attacks the immune system because the protein it uses to get inside a cell is found on certain white blood cells that are part of the immune system. When these cells are killed, the body can no longer fight off infections, making the patient highly susceptible to disease. But before this happens, the body wages a protracted war against HIV, producing antibodies to the virus. These antibodies are evidence of HIV infection and are the basis for conducting tests to see if a person has been infected—HIV-positive means a person has these antibodies. The same antibodies are also used to screen donated blood.

Vaccination helps the immune system to fight viruses because it provides a kind of trial by which the immune system learns to make antibodies to the virus. Memory cells retain this information, and when the body encounters a real infection there is no delay in the response. But no effective vaccine has been found for HIV because of its extraordinarily high mutation rates. New versions of the virus are "invisible" to the immune system until it relearns how to make new antibodies. And then, when the new antibodies begin destroying the virus, further evolution occurs and the virus becomes invisible again. The cycle repeats again and again.

HIV evolution also accounts for its origins and the initial appearance of AIDS. No one is certain where or how the virus originally evolved, but there is strong evidence pointing to a similar virus in monkeys and apes. This virus, known as simian immunodeficiency virus (SIV), infects African primates (simians) and is sexually transmitted. Some strains of SIV have a genome that shows a lot of similarity to the genome sequence of the two major types of HIV (HIV-1 and HIV-2). Although SIV does not produce AIDS in these animals, the virus can cause disease if it jumps from one species to another. The virus can jump from one species to another during instances of fighting or predation.

At some point in time, scientists believe that SIV jumped from African primates to humans, becoming HIV. Although AIDS was recognized as such only in the 1980s, physicians have tested samples from earlier patients and have found HIV in samples taken from humans as early as the 1950s. The virus could have made the transition between species much earlier, though most researchers believe that, if this theory is correct, then SIV began infecting humans some time in the early or middle of the 20th century.

How did the virus cross over to humans? There are a number of possibilities, but current evidence suggests the virus made its way into humans in the same way that it did into other species. According to this theory, known as the hunter theory, humans hunted, killed, and ate an infected animal, probably a chimpanzee.

Supporting this theory is research indicating a similar process occurring today. Nathan D. Wolfe and his colleagues at Johns Hopkins University in Baltimore, Maryland, and researchers at CDC, the Henry M. Jackson Foundation and the Walter Reed Army Institute of Research in Rockville, Maryland, and the Army Health Research Centre in Cameroon, Africa, examined blood samples from people living in rural villages in Cameroon, located in central Africa. The researchers took samples from 1,099 people who had reported contact with the blood or body fluids of primates in the wild. In a paper published in *Lancet* in 2004, "Naturally Acquired Simian Retrovirus Infections in Central African Hunters," Wolfe and his colleagues reported 10 samples that were positive for antibodies to a retrovirus, simian foamy virus, which normally infects primates. Later, Wolfe and Marcia L. Kalish at the CDC, along with many colleagues, found a case of SIV in a Cameroon hunter.

EVOLUTION AND EPIDEMICS

The transition to humans is only the first step in the process. Somehow a virus that normally thrives in a certain species must adapt itself to the new host. In most cases, a human who comes into contact with an animal virus suffers no ill effects, for the virus cannot latch onto any receptors so it does not enter, and infect, any cells. But once in a while a mutation at the right moment will occur. If there are many opportunities for the virus to infect a person, then sooner or later a mutated virus will by chance have the right proteins to gain entrance into one or more kinds of human cell and reproduce. The animal virus has now become a human one.

High mutation rates of RNA viruses such as SIV, HIV, and the various influenza strains generate large numbers of slightly different viruses. Most of these viruses fail and vanish. But when a new virus adapts to a human host, it may be able to infect other humans. This is what researchers believe happened when SIV became HIV. Although a rare event, there were probably several instances in which SIV adapted to a human host, and these different cases evolved into the different strains of HIV. HIV-1 is much more transmissible than HIV-2, so HIV-1 is more infectious and has caused many more cases of the illness than HIV-2.

Health personnel are worried that something similar may happen again. Bird influenza viruses, such as H5N1, have already jumped to humans in a few instances, although they have not evolved yet to become easily transmissible. Other recent viral infections include SARS, severe acute respiratory syndrome, which appeared in Asia in late 2002 and early 2003, and quickly spread to America and Europe. More than 8,000 people became infected, and 774 died. Researchers isolated the SARS virus in just a few months, and sequenced it a short time later. The virus is a single-stranded RNA virus, transmitted by close contact with an infected person (such as coming into contact with an infected person's sneeze or cough droplets). Isolation and treatment of infected persons halted the spread of the virus and prevented further cases.

Although government agencies succeeded in keeping SARS corralled, the ability to stop a potential epidemic depends on the ability to act faster than the virus. Viruses may spread sexually, as HIV, or by close contact, as SARS, or, in the worst case, through the air, as do the common strains of human influenza. The danger also depends on how virulent the virus is. Successful parasites such as the common human influenza viruses often do not kill their hosts—for parasites, "biting the hand" that feeds them means losing their home. If a parasite does kill its host, it cannot do so quickly, or the parasite will expire before it can infect other hosts. This is why the deadliest viruses tend to disappear rapidly. An unfortunate exception is HIV, which is one of the deadliest viruses to infect humans but kills slowly, over the course of years, which gives the virus time to find other victims.

One of the most important areas of research in virology is the study of the mutations that determine which species a virus can infect and its method of transmission. An extremely important example of such studies involves the Spanish flu and the associated virus, which took millions of lives in 1918. Because the sciences of virology and genetics were in their infancy in the early part of the 20th century, researchers at the time made little progress. Keeping the samples they acquired did not seem to be a good idea at the time because of the fear of continued spread—and with World War I fresh in people's minds, maintaining a potentially deadly biological weapon was unthinkable. Modern researchers know much more about viruses and have developed many techniques to study them safely, but finding samples of the 1918 virus and reconstructing its genomic sequence after a span of nearly a century was not a trivial task, as the following sidebar indicates.

Scientists at several laboratories have been working on the 1918 virus since its genes were identified and reconstructed by researchers at the Armed Forces Institute of Pathology in Washington, D.C. Armed Forces Institute of Pathology researchers Jeffrey Taubenberger, Ann Reid, and their colleagues have studied the three enzymes, called polymerases, the virus uses to replicate its genome. They found the gene sequences for these proteins are highly similar to avian (bird) influenza viruses, suggesting that they evolved directly from this source with a small number of mutations. If this proves to be correct, this means the 1918 virus did not arise from reassortment, which results when multiple viruses infect the same host and swap genetic segments. The researchers published their finding in 2005 in *Nature,* "Characterization of the 1918 Influenza Virus Polymerase Genes."

The 1918 Influenza Virus

A 21-year-old private in the United States Army, stationed at Camp Jackson in South Carolina, reported to the base hospital on September 20, 1918. Six days later he became one of the millions of people who died during the Spanish flu epidemic. Yet something scientifically important emerged from his tragedy nearly 80 years later, for some of his tissues had been chemically preserved and were located at the Armed Forces Institute of Pathology in Washington, D.C. In 1997, researchers Jeffrey Taubenberger, Ann Reid, and their colleagues at the institute published a report in *Science* after studying this and other preserved tissue from the time of the epidemic. As described in the paper, "Initial Genetic Characterization of the 1918 'Spanish' Influenza Virus," the researchers sequenced nine fragments of RNA from the virus.

But the entire sequence continued to elude scientists. RNA is more susceptible to degradation than DNA, so the chance of finding an intact viral genome was slim. Yet the researchers got lucky when they examined the remains of an Inuit woman who had been buried in a mass grave on the Seward Peninsula of Alaska. Preserved by permafrost—frozen ground—her lung tissue provided more viral RNA fragments. These and other fragments finally gave researchers the information they needed to reconstruct the 1918 virus's genome.

Considering the virulence of this virus, its reconstruction created some nervousness. In addition to the standard precautions virologists take when working with extremely dangerous viruses, researchers working with this virus wear face shields with air purifiers, shower when exiting the lab, and take antiviral medications. There is also some controversy about publishing the findings because some people worry that terrorists might use the information. But researchers, along with the editors of science journals, have decided the benefits from publishing the research, which makes the findings known to those who could potentially have to fight any future epidemic, outweigh the risks.

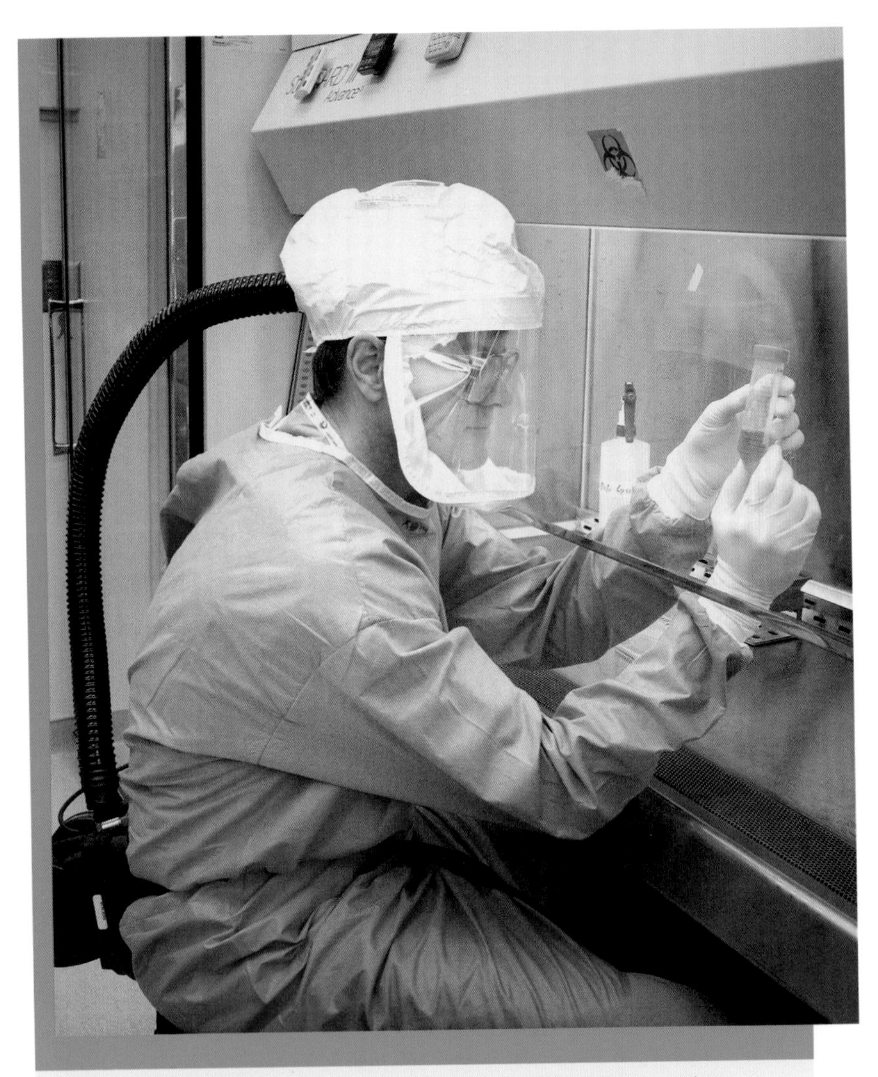

Terrence Tumpey examines a reconstructed 1918 influenza virus. *(James Gathany/CDC)*

Subsequent research has attempted to find which amino acids were mutated to produce this highly infectious virus. In 2007, Terrence M. Tumpey and his colleagues at CDC, along with researchers at Mount Sinai School of Medicine in New York and the Agricultural Research Laboratory at Athens, Georgia, deliberately mutated the sequence of the hemagglutinin gene and tested the new viruses. They found that a two-amino-acid change resulted in a drastic reduction in the transmissibility of the virus among laboratory animals. As described in the report, "A Two-Amino Acid Change in the Hemagglutinin of the 1918 Influenza Virus Abolishes Transmission," published in *Science,* the researchers hypothesize that in 1918 the change went in the opposite direction. This would have turned a difficult-to-catch virus into a highly infectious one.

Scientists do not know for certain how the 1918 influenza virus evolved, nor can they predict the evolution of H5N1 and other current viral threats. But research at this frontier of science is leading to a better understanding of how viruses change from harmless bits of nucleic acid and protein into dangerously contagious germs.

CONCLUSION

Although many biologists do not consider viruses to be alive—they are incapable of generating energy or reproducing themselves unless they infect a cell—viruses are clearly able to undergo evolution. Some viruses, especially the RNA viruses such as HIV and influenza, mutate rapidly, making it difficult for immune systems and vaccine-producing manufacturers to keep up. For a virus, a fast evolutionary track is essential. Viruses that fail to evolve quickly enough find themselves without a host. As happened to smallpox, this means such viruses disappear.

No one mourns the loss of the smallpox virus. But viruses have their brighter side, and some are immensely useful. The basic skill of any virus is to sneak its RNA or DNA inside a cell and make many copies of its genes by pirating the cell's protein machinery. Scientists working with genetic material often need these very skills. For instance, in the process of sequencing large genomes such as the human genome, scientists use enzymes to break apart the DNA. The DNA pieces can be stored for future use by inserting them in bacteria, but the pieces may also be stored in certain types of bacteriophage. DNA of these bacteria viruses can be chopped with enzymes, and then scientists employ other enzymes to insert a piece of DNA to be stored and seal the ends together. A collection of these phages that contain the entire set of DNA pieces is known as a phage library.

Researchers are also finding plenty of applications for retroviruses that make copies of their genes and insert them into a cell's genetic material. Although scientists can easily make copies of genes in the laboratory, getting the DNA inside the nucleus of a cell and integrating this DNA into the cell's genetic material is exceptionally difficult—this procedure requires the skills of a retrovirus, which have evolved to do this very operation. Of course researchers do not wish to handle dangerous viruses, so they alter the genetic material of these viruses before using them. The viruses become harmless, but retain their ability to make DNA copies of their genes and slip them into a cell's DNA.

This procedure could have enormous medical applications because not only can scientists rip out the harmful viral genes, they can also replace them with beneficial ones. A faulty gene can cause serious diseases in humans, but if physicians had a way of replacing the bad gene with a good copy, the disease could be cured. This procedure is known as gene therapy.

Although gene therapy has a lot of promise, researchers have been working on this technique since the 1980s and have not as yet made much progress. The problem is that viruses tend to insert the genes randomly in the cell's DNA, without regard to what is already there. If a virus inserts the genetic material in the middle of a gene that the cell needs, this gene may no longer function and the cell may either die or grow out of control. In the late 1990s, physicians succeeded in using gene therapy to cure a rare disease known as severe combined immunodeficiency in several young patients, but some of the patients developed leukemia.

The complexity of viruses is just beginning to be unraveled. In some of the simpler cases, researchers understand them well enough to use them in the laboratory as well as to create vaccines to stop the spread of infection. But as viruses continue to evolve and change, virologists must keep expanding the frontiers of knowledge. Avoiding another catastrophe such as the 1918 influenza epidemic requires a fast and effective response to any fresh outbreak, which demands monitoring, evaluation, and rapid characterization of the viral threat. And if scientists can achieve an even more complete understanding, viruses may even cure instead of kill.

CHRONOLOGY

1000 B.C.E. Chinese increase their resistance to smallpox by exposing themselves to a small amount of infected matter.

1796 C.E.	British physician Edward Jenner (1749–1823) develops a vaccination for smallpox.
1892	Russian researcher Dimitri Ivanowsky (1864–1920) observes that the agent causing tobacco mosaic disease is so tiny it slips through the smallest filter.
1898	Dutch botanist Martinus Beijerinck (1851–1931) studies the agent causing tobacco mosaic disease and calls it a virus.
1915	British scientist Frederick Twort (1877–1950) discovers viruses that infect bacteria. These viruses are later named bacteriophages.
1918–19	The Spanish flu pandemic claims about 50 million victims worldwide.
1933	British virologist Patrick Laidlaw (1881–1940) and his colleagues isolate the first human influenza virus.
1935	American chemist Wendell Stanley (1904–71) makes crystals with the tobacco mosaic virus, and shows that it is composed of arrangements of proteins and nucleic acids.
1945	The first influenza vaccine becomes available in the United States.
1946	The United States establishes the Communicable Disease Center, later to be known as the Centers for Disease Control and Prevention (CDC).
1957–58	An influenza outbreak begins in China and spreads, causing a pandemic that kills several million people worldwide. This pandemic is sometimes called the Asian flu.
1968	An influenza outbreak begins in Hong Kong and spreads, causing a pandemic that kills about one

million people worldwide. This pandemic is sometimes called the Hong Kong flu.

1976 Belgian biologist Walter Fiers and his colleagues sequence the complete genome of an RNA bacteriophage, MS2, the first complete genome to be sequenced.

1977 British biochemist Frederick Sanger (1918–) and his colleagues sequence the complete genome of a DNA bacteriophage, φX174, the first DNA genome to be completely sequenced.

1980 The World Health Organization (WHO) announces that global vaccination efforts have succeeded in eradicating the smallpox virus in the wild.

1981 American physicians identify cases of a new disease later to be known as acquired immunodeficiency syndrome (AIDS).

1983 American researcher Robert Gallo and French researcher Luc Montaigner discover HIV, the virus that causes AIDS.

1990s Gene therapy using a virus succeeds in curing several patients of a rare disease called severe combined immunodeficiency, but some of the patients later develop leukemia.

1997 Influenza virus H5N1, also known as bird flu, infects a human for the first time.

2002–03 Severe acute respiratory syndrome begins in Asia and spreads to other continents, killing more than 750 people. In April 2003, scientists isolate the SARS virus.

2005 Armed Forces Institute of Pathology researchers Jeffrey Taubenberger, Ann Reid, and their colleagues finish reconstructing the 1918 influenza virus.

2009 WHO officials announce that H5N1 influenza has killed more than 250 victims worldwide.

FURTHER RESOURCES

Print and Internet

American Society for Microbiology. "Meet the Microbes: Viruses." Available online. URL: http://www.microbeworld.org/microbes/virus/. Accessed April 1, 2009. The American Society for Microbiology is an organization of scientists who study microorganisms. This page describes viruses—what they are and how they work.

Barry, John M. *The Great Influenza: The Story of the Deadliest Pandemic in History.* New York: Penguin Books, 2005. The 1918 influenza pandemic swept across the globe, leaving 50 million victims and a stunned world in its wake. This book tells the story of what many people believe is the worst pandemic in history.

Cann, Alan J. *Principles of Molecular Virology,* 4th ed. London: Elsevier Academic Press, 2005. This advanced textbook describes the properties and classes of viruses.

Centers for Disease Control and Prevention. "Avian Influenza (Bird Flu)." Available online. URL: http://www.cdc.gov/flu/avian/. Accessed April 1, 2009. The Centers for Disease Control and Prevention (CDC) is a United States government agency that monitors and investigates hazards and diseases, including viral outbreaks that have not yet reached the United States but could potentially do so. This Web site discusses the current status of avian influenza virus H5N1.

———. "Disease & Conditions." Available online. URL: http://www.cdc.gov/DiseasesConditions/. Accessed April 1, 2009. Links on this page point to information on a variety of viral diseases, including influenza, HIV, herpes, and others.

Davidson, Michael W., and Florida State University. "Virus Structure." Available online. URL: http://micro.magnet.fsu.edu/cells/virus.html. Accessed April 1, 2009. This page gives a detailed description of the parts of a virus.

Davies, Pete. *The Devil's Flu: The World's Deadliest Influenza Epidemic and the Scientific Hunt for the Virus That Caused It.* New York: Holt,

2000. This book describes the 1918 influenza pandemic as well as the search by scientists to recover and study the extremely contagious virus that caused it.

Department of Health and Human Services. "Pandemic Flu." Available online. URL: http://www.pandemicflu.gov/. Accessed April 1, 2009. The United States Department of Health and Human Services maintains this Web site, which provides updated news and information concerning the potential development of pandemic influenza and the ongoing strategies to prevent its occurrence.

Drake, John W. "Rates of Spontaneous Mutation among RNA Viruses." *Proceedings of the National Academy of Sciences* 90 (1993): 4,171–4,175. Drake measured the mutation rates for influenza and HIV.

Franklin, Benjamin. *The Autobiography of Benjamin Franklin.* 1771–78. Available online. URL: http://etext.virginia.edu/toc/modeng/public/Fra2Aut.html. Accessed March 31, 2009. Written as a series of letters over the course of a number of years, Franklin's autobiography describes his many business ventures and his activity in politics.

Kanabus, Annabel, Sarah Allen, and Bonita de Boer. "The Origins of HIV and the First Cases of AIDS." Available online. URL: http://www.avert.org/origins.htm. Accessed April 1, 2009. This well written essay documents what is known and conjectured on the origin of the virus that causes AIDS.

Peters, C. J., and Mark Olshaker. *Virus Hunter.* New York: Anchor Books, 1998. The author, an expert who has spent many years studying viruses in the laboratory as well as in the field, details some of his encounters with the most interesting and often dangerous viruses in the world.

Public Broadcasting Service. "The American Experience: Influenza 1918." Available online. URL: http://www.pbs.org/wgbh/amex/influenza/index.html. Accessed April 1, 2009. As the companion Web site to the PBS program, these pages include photographs, interviews, maps, and other information describing the 1918 influenza epidemic in the United States.

Sompayrac, Lauren. *How Pathogenic Viruses Work.* Sudbury, Mass.: Jones and Bartlett Publishers, 2002. Intended for biology students, this lively book takes the reader inside the world of viruses, and explains the tricks that viruses use to take advantage of their hosts.

Taubenberger, Jeffrey K., Ann H. Reid, Amy E. Krafft, Karen E. Bijwaard, and Thomas G. Fanning. "Initial Genetic Characterization of the 1918 'Spanish' Influenza Virus." *Science* 275 (March 21, 1997): 1,793–1,796. The researchers report the sequence of nine fragments of RNA from the 1918 virus.

Taubenberger, Jeffrey K., Ann H. Reid, Raina M. Lourens, Ruixue Wang, Guozhong Jin, and Thomas G. Fanning. "Characterization of the 1918 Influenza Virus Polymerase Genes." *Nature* 437 (October 6, 2005): 889–893. The researchers studied three enzymes, called polymerases, that the 1918 virus used to replicate its genome.

Tumpey, Terrence M., Taronna R. Maines, Neal Van Hoeven, Laurel Glaser, Alicia Solórzano, Claudia Pappas, et al. "A Two-Amino Acid Change in the Hemagglutinin of the 1918 Influenza Virus Abolishes Transmission." *Science* 315 (February 2, 2007): 655–659. The researchers mutated the sequence of the gene and, after testing the new viruses, found that a two-amino-acid change resulted in a drastic reduction in the transmissibility of the virus among laboratory animals.

Wolfe, Nathan D., William M. Switzer, Jean K. Carr, Vinod B. Bhullar, Vedapuri Shanmugam, Ubald Tamoufe, et al. *Lancet* 363 (2004): 932–937. "Naturally Acquired Simian Retrovirus Infections in Central African Hunters." Wolfe and his colleagues examined blood samples from people living in rural villages in Cameroon who had reported contact with the blood or body fluids of primates in the wild. The researchers found 10 samples that were positive for antibodies to a retrovirus, simian foamy virus, which normally infects primates.

6

Regeneration—Healing by Regrowing

Until the 20th century, all sponges used for cleaning and washing came from the dried bodies of sea creatures—marine sponges. Most sponges today are made from artificial materials, but natural sponges are extremely absorbent, and many soldiers in ancient times made sure to pack a dried marine sponge so that they could take care of personal business during a long campaign. But living sponges have an even more remarkable property. If the soft body of a living sponge is disassembled and separated into small fragments—which happens if someone pushes a sponge through a net, for example—the pieces can reassemble into a living animal!

Sponges are among the simplest animals. There are a variety of species, some of which are small and some of which grow up to about 6.5 feet (2 m); most sponges live in the ocean and attach themselves to the seabed. Usually cylindrical in shape, sponges eat by drawing water into a central cavity and filtering food particles. Sponges are generally not mobile so they have no need for nerves or muscles. Most of the cells of the sponge are not specialized for any purpose at all, which makes reassembly possible. If the parts had specific places and positions, a fragmented sponge would be highly unlikely to pull itself together without someone placing the parts in the correct orientation.

Missing pieces are no problem, since sponges are capable of *regeneration*—regrowing a lost or damaged part. Such recuperative power would be immensely useful in humans as well, but unlike sponges, humans and most animals have specialized organs and tissues that make reassembly impossible. Yet certain animals can regenerate limbs and portions of the heart. This chapter describes how scientists study these amazing regenerative abilities, which may one day be applied to human patients.

INTRODUCTION

Simple organisms such as bacteria consist of a single cell. Larger and more complex organisms contain many cells, but even so, the cell is the basic unit of life for all organisms on Earth, large and small. In both single-celled and multi-cellular organisms, a cell performs the basic functions of extracting energy from nutrients, as well as generating essential proteins by expressing genes. In bacteria and other single-celled organisms, this is all that is necessary; for larger organisms, cells usually form organs or systems and perform specialized tasks. Cells that specialize are slightly different from other cells of the body, even though all cells descend from a single fertilized egg cell (the union of a sperm cell from the father and an ovum, or egg cell, from the mother). In the course of development of a multi-cellular organism, cells *differentiate.*

The human body, for example, contains many different types of cell. Although cells have a certain number of characteristics in common, cells of the brain have special properties that are not found in cells of the heart, liver, muscle, skin, blood, and other tissues and organs, and the same is true for other cells. For instance, certain cells of the brain, heart, and muscle contain proteins that are sensitive to electrical potential and allow the flow of charged particles called ions into and out of the cell. These proteins, known as ion channels, orchestrate an electrical impulse that exists only in the brain, heart, and muscle. In the brain, cells use these impulses to communicate with one another as the brain processes information, whereas heart cells specialize in the use of these impulses to contract and pump blood. Skeletal muscles also work by contracting, but they are attached to the skeleton (via tendons) and specialize in moving the body.

As discussed in chapter 3, proteins are the workhorses of the body. Some proteins, such as those involved in breaking down nutrients, are common to all cells. The proteins in brain, heart, and muscle cells that generate electrical impulses are found only in those cells. A number of proteins in other tissues and organs are also unique to certain cells, allowing them to perform their specialized tasks.

Cells make specific proteins by using, or expressing, the genes that code for these proteins. Chapter 2 described the genome, which includes the set of all genes of an organism. In humans, the genome has been sequenced and contains roughly 20,000 genes, and almost all the cells of the body contain the whole genome. Yet cells do not express the whole set of genes, they express only the subset that they need to perform their specialized function. Brain, heart, and muscle cells express the genes necessary to create the electrical impulses, along with the other genes that allow them to use these impulses in their own unique way. Other cells do not express these genes because they do not need the proteins.

Cells begin to specialize over the course of the organism's development. Some animals, such as chickens and salamanders, develop inside a protective egg, while others, such as almost all mammals, develop in the mother's womb. The starting point is the same for all animals—a large, fertilized egg cell that begins to divide. Before a cell divides it copies its DNA; when a cell divides it splits into two cells, each of which have the normal complement of chromosome pairs. (Certain reproductive cells are exceptions to this rule, producing sperm or egg cells that have only one set of chromosomes.) The original set of genes, created when the sperm and ovum united, gets handed down for all successive cellular generations.

Developing organisms, which are called embryos, become larger as the cells grow and divide. At the early stages, embryos resemble a ball of cells, but as development proceeds, cells begin to differentiate and migrate to their proper positions. Tissues and structures begin to form. In humans, most of the main structures appear by two months in development, after which time the embryo is called a fetus.

Differentiation occurs as cells switch on certain genes, making the proteins they need to specialize. The process is coordinated throughout development, as tissues organize into organs and limbs grow at the proper places. The transformation of a ball of cells into a complex organism with a brain, heart, and other complicated structures is one of the most amazing spectacles in all of science. Biologists who study development have discovered important contributions made by a few special genes that control or guide the overall structural plan, and substances that initiate development at certain points in time and at certain positions. But the study of development remains an active area of research.

Following birth, most of the major structures are in place, although babies of most species continue a slow process of growth. Adulthood marks the end of the growth phase and the beginning of maturity. Yet even in adults, cells such as those in the blood and skin die and must be replaced, injuries must heal, and proteins are constantly being turned over. A living organism is never static.

But the remarkable activities that developed new tissues and organs are no longer evident in most adult organisms. Although the brain and the immune system continue to change as they learn and remember new events, the cause of these changes is not a whole new set of tissues but rather a series of adjustments of the existing system. Adults do have the capacity to generate new tissue in some cases, but only to a limited extent and in response to specific events. A great deal of this capacity comes from important cells known as *stem cells.*

STEM CELLS

Many differentiated cells in an adult organism cannot divide. Once they have finished maturing into a differentiated cell they will not revisit their youth—they are set in their ways, so to speak. For example, skin cells form a protective covering, and red blood cells carry oxygen throughout the body, and both of these cell types have unique properties to do their job. Once they have differentiated, they no longer divide. (Red blood cells do not even have a nucleus, for it was lost in the process of differentiation. These cells are exceptional in that they do not have any of the genes that usually reside in the DNA of the cell's nucleus.)

Yet people continually lose skin and blood cells throughout their lives. Skin is exposed to the elements and eventually wears out, and blood cells die after about 120 days in circulation. These losses are in addition to any injuries that may damage the skin or cause bleeding. Replacing cells that cannot divide would seem to be impossible, and certainly would be, without the help of a pool of undifferentiated cells ready to mature and take their place. These unspecialized cells are called stem cells.

The term *stem cell* implies that other cells of the body stem from these unspecialized cells. American scientist Edmund B. Wilson (1856–1939) made the term popular in the English language after he used it in his 1896 book, *The Cell in Development and Inheritance,* though he was only translating an earlier German term. In 1868, German biologist Ernst Haeckel (1834–1919) wrote about a *stammzelle* (German for stem cell) that he believed was the ancestor cell from which other organisms evolved. The name eventually came to refer to the undifferentiated cells that give rise to other, more specialized cells.

Stem cells are able to maintain their youthful state indefinitely. Their perpetual youth contrasts with other cells. American biologist Leonard Hayflick (1928–) noticed in the 1960s that cells in culture—grown in glass dishes and provided with all the necessary nutrients—will not divide more than about 50 times before dying. These cells may be fully differentiated (yet capable of dividing) or they may be only partly differentiated, but in either case their replicating ability is limited. This limit, known as the Hayflick limit, suggests that aging is a normal and inescapable process that affects cells.

But stem cells ignore this limit. They have not yet gone very far down the path leading to maturity, so they stay in a sort of childlike innocence. Stem cells replace old or damaged tissue by dividing when necessary. Some of these offspring differentiate and assume their new duties, while other offspring remain stem cells, replenishing the stock. For example, stem cells located in the cavities of bones make red blood cells, replacing the millions that die every day. These bone marrow stem cells are called pluripotent because they can produce not only red blood cells but also other cells in the blood, such as white blood cells, as shown in the figure.

How do stem cells in the bone marrow know how many red blood cells to generate? The process is under the control of a hormone called erythropoietin (EPO), secreted by the kidney when the body's tissues are not receiving enough oxygen and need more oxygen-carrying red blood cells. (EPO is one of the substances banned by the Olympics, because unscrupulous athletes can gain an advantage by fooling their bodies into making more red blood cells, which delivers more oxygen to their muscles and increases the athlete's endurance.)

Growth is extremely important to control. Just as the development of young organisms proceeds only until its structures are finished and in the right place, replacement of tissue in adults must stop when no more is needed. Uncontrolled growth leads to a disease called cancer, of which many different varieties can afflict humans. Cancer is the second leading cause of death in the United States.

Controls on growth limit the opportunities of stem cells. Most stem cells in an adult may also be constrained in the types of cell they can produce. Hematopoietic stem cells generate blood cells but not skin

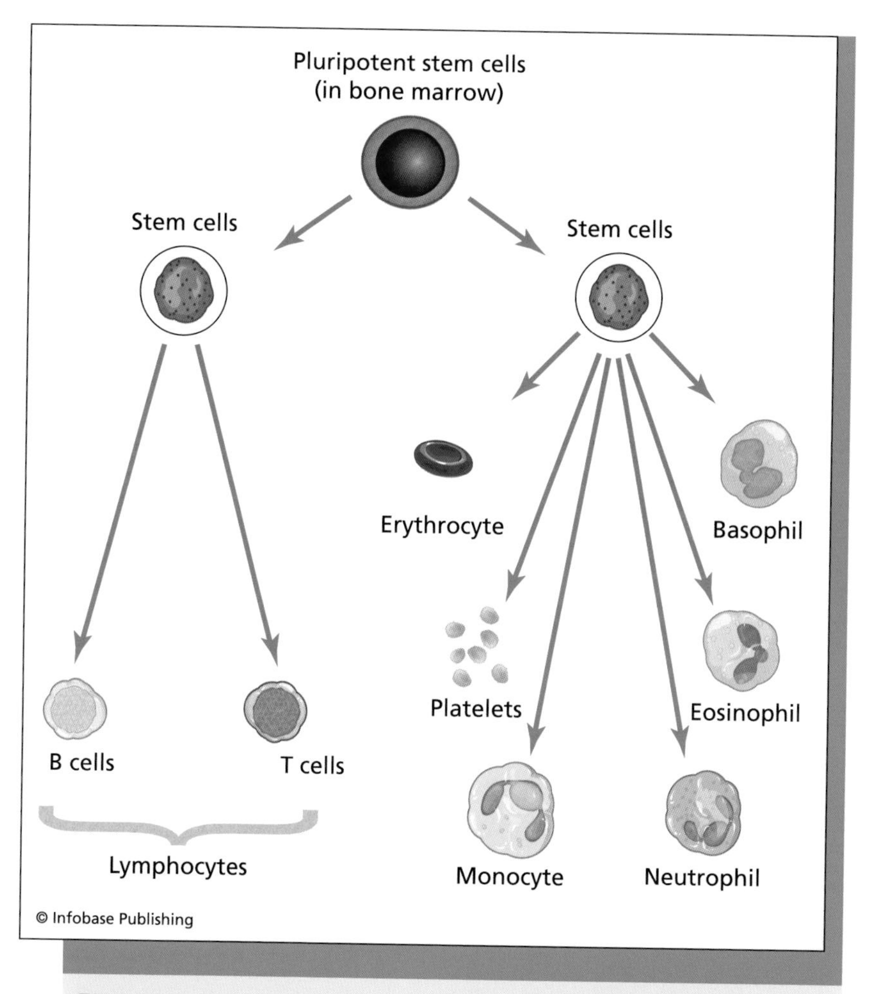

Pluripotent stem cells in the bone marrow can differentiate into many kinds of cell. Some become stem cells that will eventually differentiate into white blood cells known as lymphocytes, and others become stem cells that differentiate into red blood cells (erythrocytes) or various kinds of white blood cell.

cells, which is the job of stem cells in the lower layers of the skin. Stem cells in the brain create new brain cells, and stem cells in other tissues perform similar functions. These constraints are unlike the stem cells that appear in the initial phases of an organism's development; these embryonic stem cells are completely unspecialized and have the potential of creating any cell.

Planarians

Planarians are flatworms, named because their bodies are elongated and flattened. About 20,000 flatworm species exist, such as parasites that can grow to 65 feet (20 m) and live in the digestive tracts of animals, as well as the much smaller planarians, which include hundreds of different species that live in ponds, streams, and, much to the annoyance of tropical fish enthusiasts, aquariums. Adult planarians can reach about 0.8 inches (2 cm) in length, although these animals have an

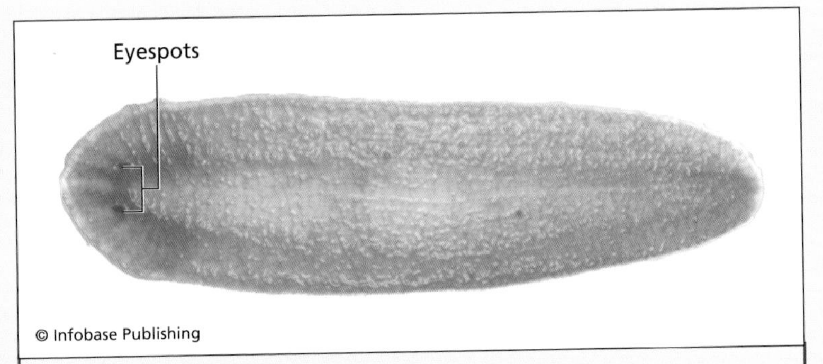

The head of a planarian has two eyespots that are sensitive to light. The body broadens in the middle and then tapers down to the tail end.

The constraints on adult stem cells suggest that they may have traveled part of the way to maturity. In other words, they have specialized a little bit—not enough to become a fully differentiated cell, but enough to constrain the kind of differentiated cell which they can produce. Yet this may not always be the case, and the versatility of adult stem cells should not be underestimated. This research is important in the effort to teach an old cell a new trick—such as regenerating a lost limb or worn-out tissues.

amazing ability to reduce their size when food is scarce. Food for a planarian consists of smaller animals or carcasses. The figure illustrates an adult planarian. *Planaria* is the name of one genus of planarians, and many people refer to planarian flatworms in general as "planaria." The name derives from a Latin word for plane, which is a flat, two-dimensional figure.

Planarians are among the simplest multicellular organisms. They do not have specialized systems for breathing or the circulation of blood, relying instead on natural movements known as diffusion to carry nutrients through their bodies. This works because planarians are flat, so the nutrients do not have to pass through many cell layers. Eyespots on the head are simple light detectors, allowing a planarian to know when it is out in the open (the animal prefers to be in the dark, under protective cover). The animals move by waving small hairs called cilia on their undersides.

The regenerative prowess of planarians has been known for a long time. German naturalist Peter Simon Sallas (1741–1811) published some notes on planarian regeneration in 1766, and biologists have been studying these animals ever since. (Early in his career, Thomas H. Morgan [1866–1945] studied planarians, but soon turned his attention to fruit flies.) If a planarian's head is cut off, for example, the tail section grows a new one. Sometimes planarians reproduce in a similar fashion, attaching their tail to a heavy object and swimming until they split into two. The result, after suitable regeneration, is two animals.

GROWING AGAIN—REGENERATION

A few species are endowed with a remarkable degree of regenerative capacity. Some of the most astonishing are flatworms known as *planarians.* The regenerative ability of a planarian has made it a common topic in science fairs at school as well as in university laboratories. The sidebar above describes these interesting animals in more detail.

Planarians can regenerate themselves with as little as 1/279th of the organism, as discovered by American biologist Thomas H. Morgan (1866–1945) in 1898. (Morgan later became more famous for his work on the genetics of fruit flies.) Somehow the animal's cells produce the right structure to replace the lost tissue—a new head or a new tail, for example. This means that the cells must have an ability to determine their position in an animal, or in other words, they must have a way of determining where they are located in relation to the head and tail. When the head is chopped off, stem cells proliferate at the site of the wound, which may be far from the tail. Yet these stem cells and their differentiated offspring realize the animal has a tail, so they appropriately regenerate the head rather than another tail. How this happens is not yet well understood, but may involve chemicals that have higher concentrations at one or the other end.

Although the positional aspect of the process is still mysterious, researchers have discovered the identity of the cells that underlie a planarian's amazing regenerative capability. Carrying out the regeneration is a large pool of stem cells called *neoblasts.* These stem cells or neoblasts are scattered throughout the body, dividing and producing new specialized cells as replacements when tissue is damaged or lost. Wounds stimulate neoblasts to move toward the site of injury and proliferate. The cells collect at a site known as the *blastema,* from which new tissue will grow. Scientists proved neoblasts are stem cells by showing that the cells are capable of creating any of the tissues of planarians.

The age of genomic science has opened a new frontier in the study of regeneration. Genes are the blueprint of an organism. The basic body plan and all of the necessary structures develop as a young organism's cells turn on and off various genes, producing the proteins that are needed as cells specialize and assume their proper shape, position, and function. By using modern genetic techniques, scientists can study the genes of these organisms with methods similar to those used to study human genes.

Phillip A. Newmark, a professor at the University of Illinois at Urbana-Champaign, and his colleagues are using these techniques to examine which genes are critical to planarian stem cells. One genetic technique involves blocking the production of specific proteins by interfering with the molecule from which the protein's amino acid sequence is constructed. (This molecule is an mRNA molecule, discussed in chapter 2.) Tingxia Guo, Newmark, and their colleagues recently reported that planarian stem cells can no longer renew themselves when a particular protein is blocked. The affected stem cells could still respond to wounds by migrating, and they begin to build new tissue, but quickly die off. Without this protein, the stem cell population fails. The research appeared in "A bruno-like Gene Is Required for Stem Cell Maintenance in Planarians," published in 2006 in *Developmental Cell.*

Other scientists are also using this method, known as RNA interference, to probe essential stem cell genes. Peter W. Reddien, Alejandro Sánchez Alvarado, and their colleagues at the University of Utah found another protein that is necessary for neoblasts to regenerate tissue. The researchers published their work in a 2005 paper in *Science,* "SMEDWI-2 Is a PIWI-Like Protein That Regulates Planarian Stem Cells."

Further research will be needed to understand what role these proteins play in stem cell biology. This research is important not only in the study of planarian regeneration but also in studies aimed at discovering why other animals, and humans, have so little regenerative capacity. Because of evolution, many genes are shared or are quite similar in organisms as different as humans and planarians. (This is why proteins and genes are sometimes described as "like" another, as in the titles of the two papers above—the researchers are pointing out a similarity to another, perhaps better known protein or gene in other organisms.) What researchers learn about planarians can be applied to other organisms, including humans. R. John Davenport, an editor for *Science,* noted in an article published July 1, 2005, in the magazine that if only a few genes are involved, "perhaps altering a handful of genes would be enough to turn us into superhealers, too."

Planarians are therefore important "model" organisms. The term *model* in this usage refers to a simple organism that displays properties researchers wish to study in more complex organisms. As in engineering and other fields of science, the model is a simpler version of a complicated object or process, although biologists usually have greater uncertainty whether their models behave in the same way as the more complex organisms.

The advantages of using planarians as models are their astonishing regenerative capacity, their small size and simple structures, and the short times required for various processes such as regeneration. But there are disadvantages as well. The uncertainty that biologists have about the applicability of model organisms increases with evolutionary distance—although all organisms, including planarians and humans, share genetic features, planarians and humans have been evolving separately for such a long time that profound and fundamental differences can arise. Closer organisms such as chimpanzees and humans share many more genes and genetic processes; chimpanzee DNA sequences, for example, differ from humans by only a few percent.

Picking a model organism that is closer to humans gives researchers a little more confidence that the research applies to humans. Although planarian research is also interesting in its own right, studies of vertebrates (animals with backbones) might offer more clues in regards to human biology.

Unlike planarians, however, most vertebrates have few regenerative capabilities. But there are exceptions. Some lizards can regrow a missing tail, which is often lost in the jaws of a disappointed predator (or to the fingers of an inquisitive child), allowing the lizard—or at least the most vital parts of it—to escape.

Zebrafish are another fascinating exception. A zebrafish is a small tropical fish that gets its name from the stripes on the side of its body, and is popular in aquariums as well as research laboratories. Scientists are interested in zebrafish because they are an excellent model organism for many experiments in genetics, particularly because their transparent embryos are easy to study. Zebrafish are also popular with scientists who study regeneration because these fish are able to regrow lost fins, as well as a portion of their heart. Regeneration in zebrafish will be discussed further in a later section of this chapter.

Perhaps the champion regenerator in vertebrates is the salamander. Salamanders belong to the classification order Urodela, which is derived from a Greek term that means having a tail. Salamanders are amphibians. Unlike other amphibians such as frogs, which have tails only when they are young (in the tadpole stage), salamanders keep their tails as adults, a feature that gives them their scientific name. There are about 500 species of salamander, including newts, which are salamanders that tend to live mostly in the water. Salamanders can regenerate not only a lost tail but also missing legs, and the animals maintain this ability throughout their lifetime. These animals can also regrow jaws, parts of the eye, portions of the brain and spinal cord, the intestine, and even small regions of the heart.

This salamander, an axolotl, lives in Mexico. *(Nature's Images/Photo Researchers, Inc.)*

The study of salamander regeneration is important in and of itself, as is research on planarians, because these intriguing vertebrates are marvels of biology. Just as importantly, salamanders offer insights that may one day lead to the development of regenerative medicine in humans.

LIMITATIONS OF TISSUE REGENERATION IN HUMANS

Unlike the amazing salamander, people do not regrow lost legs or arms. But humans are not completely bereft of regenerative ability. Wounds heal, skin repairs itself to a certain extent, and broken bones mend. In addition, young children sometimes regrow lost fingertips, though this capacity fades quickly with age. The liver also shows a remarkable ability to regenerate as much as two-thirds of the organ. This regeneration occurs, for example, after surgeons remove liver tissue because of an injury or disease. But chronic diseases such as cirrhosis, which may occur as a result of years of alcohol abuse or from other causes that continually attack the liver, are not reparable in this fashion.

Another regenerative example in humans is peripheral nerve regrowth. Nerves conduct messages from the brain to initiate and instruct the movement of muscles, as well as messages from sensory organs in or below the skin that let the brain know about touch, pressure, and pain. These nerves are called peripheral because they are outside the central nervous system (which consists of the brain and spinal cord). Damaged nerves can regrow. For example, 10 years ago the author of this book was in an accident that damaged one of his cranial nerves, which carry messages between the brain and the muscles or sensory organs of the head and face. The injury caused a loss of sensation and a paralysis (loss of movement) in the left side of the face. Besides an eerie lack of sensation, these deficits gave the author a lopsided smile and made shaving a risky endeavor. But gradually a tingling sensation returned, as did a limited ability to move. A few months later, sensation and mobility of the affected areas were nearly normal.

Such regrowth should not be confused with operations that repair damage by implanting or transplanting new tissue. Organ transplants and skin grafts help to heal serious injuries, but these processes are mediated by physicians who replace lost or damaged tissue with healthy tissue taken from elsewhere. This new tissue is not regenerated by the body; it comes from an organ donor, or from other parts of the patient's body, such as when healthy skin is removed from an unaffected area and attached, or grafted, to the site of injury. Transplants and grafts are often successful, but require donors and involve elaborate procedures. In addition, if the tissue came from another person, it may be recognized as foreign and attacked by the recipient's immune system. Patients receiving transplants typically receive drugs to decrease the activity of their immune systems, but this also raises the risk of infection.

Severed peripheral nerves regrow, though slowly, and it takes time for them to find the right target. Messages carried by the nerve are scrambled for a while until the nerve and brain adjust the connections, but eventually the nerve's function returns. Unlike peripheral nerves, though, a severed spinal cord cannot repair itself, even after a long time. Instead, in most cases scar tissue forms, impeding any chance of recovery. Scar tissue consists of connective fibers with which the body closes the wound, and often appears pale because it usually has little blood flowing to it. This tissue can appear at any site of serious injury, inside the body or on the surface. Although it takes the place of damaged tissue, scar tissue does not perform the same tasks—scars are not the result of regeneration of the same tissue that was lost, they are the body's way of keeping itself together and intact.

Limitations of adult human tissue to regenerate are in distinct contrast to the capacity of human embryos. Early in development, embryos can recover from drastic damage—for instance, after the fertilized egg cell divides and the embryo consists of eight cells, one cell can be removed with no ill effects on subsequent development. (Such removals sometimes occur during testing of embryos that are transferred to the womb during fertility treatments.)

Adult human regenerative abilities are also poor when compared to that of salamanders. This is one of the reasons why regeneration in salamanders is an important frontier of scientific research.

Limb regeneration in salamanders begins with the formation of a blastema at the tip of the injury. The limb grows from the proliferating cells of the blastema, forming a new limb in a manner similar to embryonic development. It is a self-sufficient process—if a blastema is transplanted to another part of the body, a new limb still develops. But whereas the neoblasts of planarians come from an available pool of stem cells, the proliferating cells in the growing stump of a salamander limb arise from an unexpected source—cells that have already differentiated. To regrow the limb, the cells at the wound site must undergo a process known as dedifferentiation in which they regress to an unspecialized state. This is the reverse of differentiation, which is highly unusual since differentiation is, like aging, normally a one-way process from youth to maturity. Somehow the cells are reprogrammed to enter a youthful, undifferentiated state, and then proliferate to produce the needed tissue. As discussed below, scientists are eagerly investigating this "fountain of youth" for several reasons: The mechanism is a fascinating biological process, and there is a hope that some of the principles can be applied one day to human medical procedures.

EXPERIMENTAL APPROACHES TO SPARK REGENERATION

No one is certain why salamanders and a few other animals have such advanced regenerative capacity. A possible reason why humans have little regenerative ability is related to cancer, a disease that is caused by uncontrolled growth.

Cell growth and division in most animals are highly regulated because of the threat of cancer. For instance, replacement of lost skin or blood requires new cells, but the stem cells and associated mechanisms that are responsible for this are under control so that they do not create more tissue than needed. These controls include genes that get switched on if something goes wrong, killing the cell. What happens in cancer is that a series of mutations in a cell somewhere in the body allow that cell to escape its restraints. The cell begins to divide rapidly. Because the defect is genetic, it is passed along to the progeny, so the new cells also divide uncontrollably. The result is a tumor or abnormal growth that impairs the function of the body, sometimes in life threatening ways.

Dedifferentiation occurs only rarely in humans, unlike the situation in salamander regenerative processes. Many replacement cells in the human body come from stem cells, but in the case of liver and peripheral nerve generation, some human cells seem to travel back to a youthful state temporarily. Liver cells normally do not divide often, but when regenerating lost tissue, some of them express genes that they have not switched on since the fetal stage.

Some researchers are trying to find the proteins associated with dedifferentiation processes. Other researchers are examining chemicals that may trigger dedifferentiation. Shuibing Chen, Sheng Ding, Peter G. Schultz, and their colleagues at the Scripps Research Institute in La Jolla, California, have discovered a small molecule they named reversine that reverses the course of cellular aging in mouse muscle cells—applying reversine turned differentiated muscle cells into stem cells. The researchers found that the molecule is also active in other cells. As reported in the 2007 paper "Reversine Increases the Plasticity of Lineage-Committed Mammalian Cells," published in the *Proceedings of the National Academy of Sciences,* the mechanism of reversine includes inhibiting certain proteins in cells that have already committed to a lineage (in other words, cells that have started to differentiate).

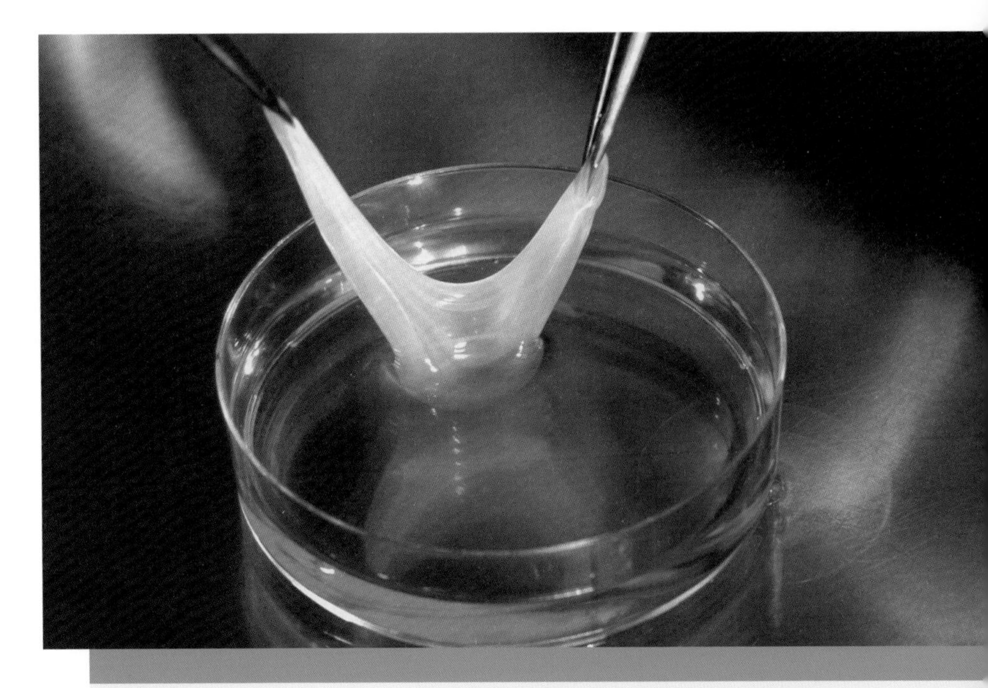

Many kinds of cells, such as the skin cells shown here, can be grown in the laboratory. *(Burger/Phanie/Photo Researchers, Inc.)*

Ding and his colleagues are continuing to study reversine, and are also searching for other factors that govern or initiate dedifferentiation. The ability to initiate and control dedifferentiation would be a significant step toward regenerative medicine in humans. Making stem cells out of older, differentiated cells bypasses the need for a stem cell pool to begin with, allowing regeneration to proceed from whatever cells are available. This project is only one of many biomedical research programs at the Scripps Research Institute. The following sidebar provides more information on this research center.

While Ding and his colleagues are searching for chemical agents, other researchers are using electricity and the flow of charged particles called ions. Ions come from the dissociation of salts such as sodium chloride (table salt), which in water splits into a positively charged sodium ion and a negatively charged chloride ion. A flow of ions constitutes an electrical current. Such currents are critical for brain function as well as playing other roles in physiology.

The Scripps Research Institute

Born in London, England, on October 18, 1836, Ellen Browning Scripps (1836–1932) and her family immigrated to the United States eight years later. An avid reader, she and her brothers founded a number of successful newspapers. Their fortunes grew. Ellen Scripps retired and moved to La Jolla, California, in 1896, but continued to be involved in the community, funding schools, libraries, and playgrounds. In 1924, she founded Scripps Memorial Hospital and Metabolic Clinic. From this institution emerged Scripps Research Institute in 1961, one of the largest nonprofit research organizations in the United States. Ellen Scripps died on August 3, 1932, in La Jolla.

Scripps Research Institute currently owns 14 laboratory buildings in La Jolla, with another small facility in Florida. More than 270 professors and hundreds of students and technicians conduct biomedical research on issues in immunology, biochemistry, neuroscience, virology, cell biology, heart disease, and other subjects. Much of the funding for this research comes from competitive grants earned by Scripps professors from the National Institutes of Health.

The philosophy of Scripps Research Institute encourages a collaborative atmosphere. This is particularly important because the solution to scientific questions often engages knowledge from a wide variety of fields. For example, research on regeneration involves principles and techniques from chemistry, animal and human biology, biophysics, and genetics. Students pursuing a Ph.D. in the biological or biomedical sciences go to Scripps Research Institute or similar institutions and universities to receive broad training in chemistry and biology.

Dany S. Adams, Alessio Masi, and Michael Levin, of Forsyth Institute and Harvard University, examined the role of ion flow in regeneration in the African clawed toad, *Xenopus.* This toad is another model organism that has been extensively studied, and Levin and his colleagues use it for regeneration research. During a period of development, *Xenopus* tadpoles can regenerate their tails, including the nerves, muscle tissue, skin, and blood vessels. Levin and his team discovered that a protein that pumps hydrogen ions (H^+) is critical for this regeneration. Regeneration fails without the hydrogen pump, even in the developmental period in which it should occur. The pump is therefore necessary for regeneration.

The researchers also found that by manipulating *Xenopus* to make a related protein, which is also a hydrogen pump, a tadpole can regenerate its tail during a time in development that the animal cannot normally do so. What this experiment shows is that the hydrogen pump is sufficient for regeneration—the process occurs when the pump is working. The hydrogen pump is therefore necessary and sufficient for regeneration in these animals. In 2007 Levin and his colleagues published this research in *Development* as "H^+ Pump-dependent Changes in Membrane Voltage Are an Early Mechanism Necessary and Sufficient to Induce *Xenopus* Tail Regeneration." Concerning the potential applications, the authors wrote, "Genetic modulation of ion flows in existing cells within wounds may be exploited by future biomedical efforts and may be a promising modality for augmenting regeneration and minimizing side effects in clinical settings."

Chemicals and ions can spark regeneration in cells that may not be normally inclined to do so, at least in these model organisms. But there is another issue to be addressed before any such method can succeed in human regenerative medicine—dealing with the scar tissue that replaced the damaged tissue. Formation of scar tissue is a quick and easy method to prevent infection that would otherwise invade wounds and make a bad situation even worse. But scar tissue not only blocks the pathway of invading microorganisms, it also tends to clog up any chance of cells to regenerate the lost tissue. This does not happen in salamanders—cells respond to injury by regenerating lost tissue rather than producing a scar. Why and how this occurs are crucial questions to answer as physicians try to induce a little more regenerative capacity in their human patients.

CARDIAC TISSUE REGENERATION

The leading cause of death in adults in the United States is cardiovascular disease—diseases of the heart and circulatory system. More than a million Americans suffer a heart attack each year. Heart attacks are caused by a loss or disruption of blood flow to the heart, leading to the death of heart cells due to a lack of oxygen. Coronary arteries carry blood to the heart, and if substances such as certain fats build up inside the vessel, the cardiac region supplied by that artery will be affected.

Blockages affecting a lot of heart tissue usually result in the patient's death because the heart stops beating. Smaller injuries kill a certain amount of tissue, but the heart begins to heal and replace the damage. As in other wounds, though, the replacement is scar tissue, which does not participate in the heart's job of pumping blood as well as the original tissue. The result is a dangerously weakened organ.

The zebrafish heart's response to injury is much different. As reported by Kenneth D. Poss, Lindsay G. Wilson, and Mark T. Keating at Harvard Medical School in 2002, one- or two-year-old zebrafish can repair a loss of 20 percent of the heart within a couple of months. Heart regeneration proceeds rapidly as large numbers of new cardiac cells cluster around the site of the wound. The researchers found a specific enzyme that was essential for this process, because when the enzyme protein was altered by a mutation, heart regeneration failed and instead scar tissue formed. These experiments were published in a paper in *Science*, "Heart Regeneration in Zebrafish."

Where did the new cardiac cells come from? Alexandra Lepilina at Duke University in Durham, North Carolina, made further investigations, along with Poss—who has moved to Duke University—and C. Geoffrey Burns of Massachusetts General Hospital in Charlestown. As reported in "A Dynamic Epicardial Injury Response Supports Progenitor Cell Activity during Zebrafish Heart Regeneration," published in *Cell* in 2006, the researchers discovered that the process begins with a blastema. The blastema consists of stem cells that differentiate into heart cells. Surrounding the injured heart is a covering whose cells migrate and form new vessels to supply the regenerating tissue. Certain proteins called growth factors stimulate the regeneration process. Unlike limb regeneration in salamanders, Poss and his colleagues found no evidence of dedifferentiation. Cells composing the blastema apparently came from a proliferation of some small reserve of stem cells.

Why zebrafish have this rare ability is not known. Mending a broken heart is not that easy in humans, yet stem cells do exist in most human tissues, including cardiac tissue. This regenerative capacity is present, in a rudimentary form, but it is not normally strong enough to regenerate fully functional cardiac tissue to replace sudden and serious losses from heart attacks.

But perhaps human regenerative power can be stimulated to achieve what zebrafish do naturally. Scientists at Hydra Biosciences, a young company in Cambridge, Massachusetts, are developing a variety of drugs to treat cardiovascular diseases, among other disorders. Some of the drugs the company is researching are proteins that elicit the proliferation of stem cells or the dedifferentiation of old cells into a youthful, unspecialized state. With these drugs, the researchers hope to create an environment that nurtures regeneration instead of stifling it with scar tissue formation. The idea is that patients would take these drugs shortly after suffering a heart attack.

Much more research recently has gone into the possibility of using stem cells obtained outside of the body for healing purposes. The concept is similar to regeneration, except the replacement tissue would derive from external sources. Embryonic stem cells have the greatest potential because they can become any cell, but this research is controversial because procuring these cells generally results in destroying human embryos. Although some people strongly support this research because of its possible benefits, other people regard the destruction of human life even in embryonic stage as immoral.

The use of adult stem cells avoids the controversies concerning embryos, but adult stem cells are usually limited by their partial maturity. Researchers are working on methods of revitalizing these stem cells and transplanting them into a patient's site of injury, but perhaps the best option is to use what is already in the patients. This technique avoids a transplantation operation as well as the risk of infection and immune system rejection that go along with it. Although success may be years in the future—and it may never come at all—a drug that effectively stimulates cardiac regeneration would save a lot of lives.

NEURAL AND SPINAL CORD REGENERATION

The central nervous system—the brain and spinal cord—is another part of the body that, if injured, often results in death or severe disabilities.

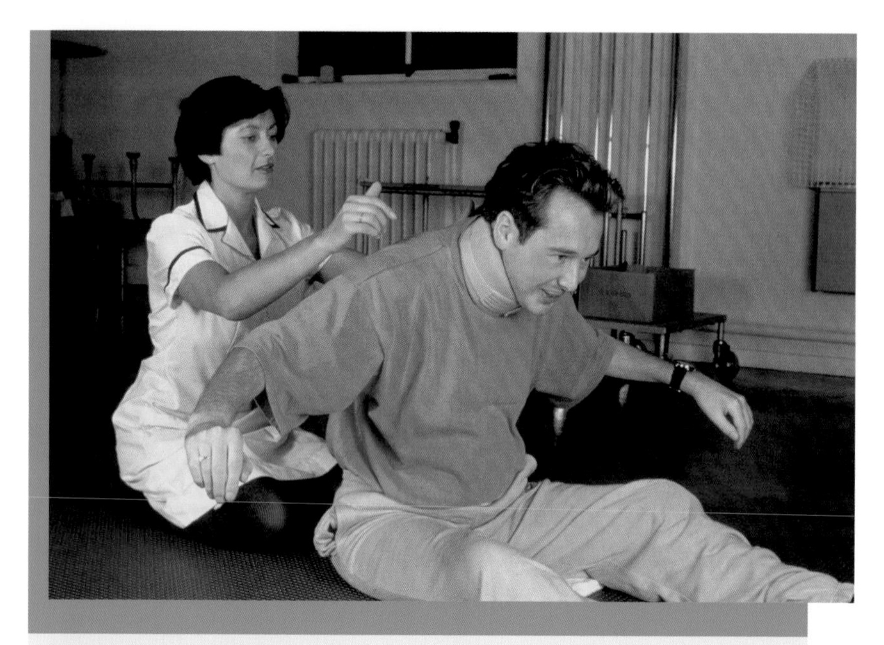

Spinal injuries can result in the loss of motor and sensory function in the parts of the body served by the damaged region. *(Simon Fraser/ Hexham General Hospital/Photo Researchers, Inc.)*

As with heart attacks, a sudden disruption of the blood supply to the brain causes cell death and can lead to paralysis (an inability to move), aphasia (difficulties in speaking and understanding language), and other serious deficits. Such incidents are called cerebrovascular accidents, but are also commonly known as strokes.

Other nervous system injuries occur when the spinal cord is damaged. The spinal cord is a bundle of nerves that passes through the backbone, carrying messages from the brain to the muscles, and from sensory cells conveying touch and pressure information to the brain. Severing the spinal cord eliminates this communication pathway. When this occurs, the brain can no longer activate the muscles, resulting in immobility, and the sensory organs cannot report to the brain, causing an absence of sensation.

Peripheral nerves are able to regenerate, as mentioned above. A cut peripheral nerve regrows because certain cells, called Schwann cells, partially dedifferentiate and guide the growing nerve back to the target. Researchers are not sure how or why Schwann cells can dedifferentiate, but it appears that cells in the central nervous system are unable to do so. As a result, a damaged spinal cord cannot regenerate. Cells in a spinal cord that has been cut, or more likely crushed, which may occur in a fall or traffic accident, will die, and are replaced with scar tissue. If some nerves of the cord remain undamaged they may be able to perform some of the functions of the damaged part, especially in young people whose bodies are more adaptable to change. But otherwise, the patient may spend the rest of his or her life confined to a wheelchair.

Paralysis and spinal cord injuries received a lot of publicity when popular actor Christopher Reeve, who had played Superman in a series of movies, was thrown from a horse on May 27, 1995, and suffered a paralyzing injury to his spinal cord. This incident galvanized research on spinal cord repair. Reeve strongly promoted this research until he died on October 10, 2004. Various strategies to repair damaged spinal cords include bridging the gap with transplanted stem cells and using electrical devices to carry signals between the brain and muscles. Success has been limited thus far.

Another strategy to repair the damage is to stimulate regeneration. Stem cells can be found in many regions of the body, although usually in small quantities. If stimulated to proliferate, stem cells in the central nervous system may be able to overcome the inhibitions placed on their growth by scar tissue and other factors. Researchers Romana Vavrek and Karim Fouad and their colleagues at the University of Alberta in Canada, along with colleagues in the University of British Columbia in Canada, have used laboratory rats as a model organism to study spinal cord injuries. These researchers have found that applying a certain protein called brain-derived neurotrophic factor (BDNF), known to stimulate nerve growth, helps the spinal cord to recover. As reported in "BDNF Promotes Connections of Corticospinal Neurons onto Spared Descending Interneurons in Spinal Cord Injured Rats," published in *Brain* in 2006, the scientists found increased sprouting of the remaining nerves. These treatments may provide a more conducive environment for healing, possibly aiding regeneration or at least allowing it to proceed as it does in the peripheral nervous system.

But one of the biggest problems is that cells in the brain and spinal cord make connections with one another that are difficult to reestablish. As discussed in chapter 1, brain cells called neurons send projections to hundreds or even thousands of other neurons and make synapses, across which they transmit information. These connections change slightly as the brain learns and remembers. Even if the brain could replace a neuron after it died, there would seem to be little chance that the new neuron could find all the old neuron's targets. Since the old neuron's function would be lost anyway, scientists used to think the adult human brain did not generate any new neurons.

Yet scientists have recently discovered that this is not true. Early reports in the 1960s and 1970s by American researchers Joseph Altman and Michael Kaplan did not receive widespread attention at the time, but their findings were later replicated. Neurogenesis—the birth of new neurons—does occur in adults. Although neurogenesis seems to be rare, it opens exciting new prospects in future treatments for brain injuries.

Neurogenesis may even be important for a number of common treatments given today. Some antidepressant drugs, used to treat depression, elevate rates of neurogenesis in an important region of the brain called the hippocampus. Although no one is certain that increased neurogenesis is the action by which these drugs work, a team of scientists including L. Santarelli and led by Reni Hen at Columbia University in New York used mice as model organisms to discover that antidepressants failed to work if neurogenesis was blocked. This report, "Requirement of Hippocampal Neurogenesis for the Behavioral Effects of Antidepressants," was published in *Science* in 2003. Whether this finding holds true in humans remains to be seen.

If physicians can find a way of stimulating neurogenesis over widespread regions of the brain, the ravages due to strokes as well as neuron-killing diseases such as Parkinson's and Alzheimer's diseases could be treated more effectively than they are today. One of the difficulties, however, is that the distribution of adult stem cells that can produce neurons seems to be restricted to a few areas.

Yet the prominent dedifferentiation that occurs in salamanders gives hope for a similar potential in humans, concealed though it may be now. Support cells in the brain known as glia can be transformed into full-fledged neurons by forced expression of certain genes, as demonstrated in 2007 by Benedikt Berninger and Magdalena Götz at Ludwig-Maximilian University of Munich, Germany, and their colleagues. The researchers showed the reprogrammed cells, cultured in dishes (*in vitro*), have properties associated with neurons. The paper appeared in *The Journal of Neuroscience*, titled "Functional Properties of Neurons Derived from *In Vitro* Reprogrammed Postnatal Astroglia."

Humans lack the regenerative capacity of planarians and salamanders, but not necessarily the potential. As researchers learn more about the mechanisms by which stem cells proliferate and by which old cells learn to become young again or get transformed into something else, the power to self-repair may grow without limit.

CONCLUSION

Proliferation and transformation are keys to regeneration. In many cases, stem cells divide rapidly to make the new cells that will compose replacement tissue, as occurs in planarians and zebrafish; in other cases, repairs come from a transformation of cells, such as the dedifferentiation of salamanders.

Regenerative medicine in humans seeks to take advantage of the latent ability of the body to heal itself. Stem cell reservoirs are sprinkled throughout the human body, and perhaps these cells could be encouraged by the application of certain factors or genes into a greater response than normal. Cells may also be nudged into expressing the proteins that would enable them to engage in functions they would not otherwise perform.

Yet if researchers succeed in mimicking the regenerative prowess of a salamander or zebrafish in a human patient, other questions will arise. Regenerating a damaged spinal cord or an injured heart would be terrific, but no one can be certain at present if the regenerated organs would do their jobs correctly or interact appropriately with other systems. Another concern involves keeping the regenerative mechanisms under control so that cancerous growths are not generated along with the desired tissues. The outcome of these issues will be critical in deciding if the most ambitious goals of regenerative medicine can ever become a reality.

Some scientists are working on an option that combines regeneration with transplantation, the procedure commonly used to replace damaged organs with those donated by other people. The idea is to grow organs in a laboratory setting, and transplant them into patients as needed. Generating new organs in a laboratory obviates the need for donations, which are often in short supply. For example, the waiting list for heart transplants is long and some patients die before a suitable organ becomes available—which occurs only as the result of another person's tragedy.

Replacing organs with laboratory-created substitutes is one of the goals of tissue engineering. A variety of methods are possible, including techniques involving mechanical devices, but many tissue engineers work with biological material. For instance, researchers at the Artificial Heart Laboratory at the University of Michigan are working on methods of layering or embedding heart cells in a matrix, producing viable heart tissue capable of supporting contraction and the pumping of blood.

The ultimate result of all of these research avenues will be a greater understanding of how cells organize themselves into tissues and organs. All organisms perform these feats once, during development, but a few organisms manage to do so repeatedly. Regeneration is a frontier at the crossroads of science and medicine that promises insight into one of the deepest secrets of nature.

CHRONOLOGY

1744 C.E. Swiss naturalist Abraham Trembley (1710–84) publishes his investigations of regeneration in a hydra, a simple freshwater organism.

1766 German naturalist Peter Simon Pallas (1741–1811) publishes the first description of regeneration in planarians.

1768 Italian researcher Lazzaro Spallanzani (1729–99) describes limb regeneration in salamanders.

1868 German biologist Ernst Haeckel (1834–1919) discusses his ideas on evolution in a German-language publication, where he uses the term *stammzelle* (stem cell) to denote the cell from which all other multicellular organisms evolve.

1890 German physician Emil Ponfick (1844–1913) describes liver regeneration in experimental animals.

1896	American scientist Edmund B. Wilson (1856–1939) uses the term *stem cell* in his book, *The Cell in Development and Inheritance.*
1898	American biologist Thomas H. Morgan (1866–1945) reports that planarians can regenerate themselves with as little as 1/279th of their body.
1909	Russian scientist Alexander Maximow (1874–1928) introduces the concept of the stem cell as a cell that gives rise to more specialized cells in the body.
1975	British biochemist Frederick Sanger (1918–) develops a method of sequencing DNA that is often used in studies involving genetics and genomes, including regeneration research.
1990s	Elizabeth Gould and her colleagues conduct experiments demonstrating neurogenesis in rats and monkeys.
2002	Kenneth D. Poss, Lindsay G. Wilson, and Mark T. Keating report that zebrafish have the capacity to regenerate their hearts following a loss of 20 percent of the tissue.
2009	President Barack Obama lifts certain restrictions on government funding of embryonic stem cell research.

FURTHER RESOURCES

Print and Internet

Adams, Dany S., Alessio Masi, and Michael Levin. "H^+ Pump-dependent Changes in Membrane Voltage Are an Early Mechanism Necessary and Sufficient to Induce *Xenopus* Tail Regeneration." *Development* 134 (2007): 1,323–1,335. The researchers found that by inducing *Xenopus* to make a particular protein, a tadpole of this species can regenerate its tail during a time in development when the animal cannot normally do so.

Bellomo, Michael. *The Stem Cell Divide: The Facts, the Fiction, and the Fear Driving the Greatest Scientific, Political and Religious Debate of Our Time.* New York: AMACOM, 2006. Although the title seems loaded with sensationalism and hyperbole, this book is an enlightening discussion of the debate surrounding the research and potential medical uses of stem cells.

Berninger, Benedikt, Marcos R. Costa, Ursula Koch, Timm Schroeder, Bernd Dutor, Benedikt Grothe, and Magdalena Götz. "Functional Properties of Neurons Derived from *In Vitro* Reprogrammed Postnatal Astroglia." *The Journal of Neuroscience* 27 (August 8, 2007): 8,654–8,664. The researchers showed that reprogrammed glial cells have properties associated with neurons.

Chen, Shuibing, Shinichi Takanashi, Qisheng Zhang, Wen Xiong, Shoutian Zhu, Eric C. Peters, Sheng Ding, and Peter G. Schultz. "Reversine Increases the Plasticity of Lineage-committed Mammalian Cells." *Proceedings of the National Academy of Sciences* 104 (June 19, 2007): 10,482–10,487. The researchers have discovered a small molecule they named reversine, which reverses the course of cellular aging in mouse muscle cells.

Coen, Enrico. *The Art of Genes: How Organisms Make Themselves.* Oxford: Oxford University Press, 1999. Developmental biology is a subject that can rapidly overwhelm a casual reader with a flood of indecipherable terms, but this book describes, in a manner accessible to students, how plants and animals develop.

Davenport, R. John. "What Controls Organ Regeneration?" *Science* 309 (July 1, 2005): 84. This single-page article summarizes important research issues in organ regeneration.

Gardiner, David M., and Susan V. Bryant. "Limb Regeneration Laboratory." Available online. URL: http://regeneration.bio.uci.edu/. Accessed April 1, 2009. These University of California, Irvine, researchers study salamander regeneration. Their well-organized and informative Web site explains the basics of regeneration and describes the laboratory's current projects.

Guo, Tingxia, Antoine H. F. M. Peters, and Phillip A. Newmark. "A bruno-like Gene Is Required for Stem Cell Maintenance in Planarians." *Developmental Cell* 11 (2006): 159–169. The researchers report that planarian stem cells can no longer renew themselves when a particular protein is blocked.

King, Rita Mary. *Biology Made Simple.* New York: Broadway Books, 2003. Part of the "Made Simple" set of volumes, this book is intended for students and breaks down a large and complex subject into more comprehensible parts. Terms and concepts are concisely explained.

Lepilina, Alexandra, Ashley N. Coon, Kazu Kikuchi, Jennifer E. Holdway, Richard W. Roberts, C. Geoffrey Burns, and Kenneth D. Poss. "A Dynamic Epicardial Injury Response Supports Progenitor Cell Activity during Zebrafish Heart Regeneration." *Cell* 127 (2006): 607–619. The researchers discovered that the heart-healing process in zebrafish begins with a blastema.

McGowan Institute for Regenerative Medicine. "Regenerative Medicine." Available online. URL: http://regenerativemedicine.net/. Accessed April 1, 2009. The McGowan Institute for Regenerative Medicine, established by the University of Pittsburgh, maintains this Web resource full of news and information on the rapidly developing field of regenerative medicine.

National Institutes of Health. "Stem Cell Information." Available online. URL: http://stemcells.nih.gov/. Accessed April 1, 2009. NIH, the main United States government agency that funds biomedical research, maintains an informative Web resource devoted to stem cells. Everything from basic science to medical applications is covered.

Newmark Lab. "Planarian Regeneration." Available online. URL: http://www.life.uiuc.edu/newmark/. Accessed April 1, 2009. Phillip A. Newmark is a professor at the University of Illinois at Urbana-Champaign. This Web page describes his laboratory's research into the amazingly regenerative planarians.

Panno, Joseph. *Stem Cell Research: Medical Applications and Ethical Controversy.* New York: Checkmark Books, 2006. Aimed at students and young adults, this book accurately explains recent research on stem cells and the controversies such research has produced, and how these cells may one day be used to treat a large number of serious diseases.

Poss, Kenneth D., Lindsay G. Wilson, and Mark T. Keating. "Heart Regeneration in Zebrafish." *Science* 298 (December 13, 2002):

2,188–2,190. The researchers report that one- or two-year-old zebrafish can repair a loss of 20 percent of the heart within a couple of months.

Reddien, Peter W., Néstor J. Oviedo, Joya R. Jennings, James C. Jenkin, and Alejandro Sánchez Alvarado. "SMEDWI-2 Is a PIWI-like Protein That Regulates Planarian Stem Cells." *Science* 310 (November 25, 2005): 1,327–1,330. The researchers found a protein that is necessary for neoblasts to regenerate tissue.

Santarelli, L., M. Saxe, C. Gross, A. Surget, F. Battaglia, S. Dulawa, et al. "Requirement of Hippocampal Neurogenesis for the Behavioral Effects of Antidepressants." *Science* 301 (August 8, 2003): 805–809. The researchers report that antidepressants fail to work in a mouse model of depression if neurogenesis is blocked.

University of Utah. "Stem Cells in the Spotlight." Available online. URL: http://learn.genetics.utah.edu/units/stemcells/. Accessed April 1, 2009. Presented by the Genetic Science Learning Center at the University of Utah, this Web resource provides an entertaining series of animated tutorials on stem cell research and possible stem cell therapies.

Vavrek, R., J. Girgis, W. Tetzlaff, G. W. Hiebert, and K. Fouad. "BDNF Promotes Connections of Corticospinal Neurons onto Spared Descending Interneurons in Spinal Cord Injured Rats." *Brain* 129 (2006): 1,534–1,545. These researchers have found that applying a certain protein called brain-derived neurotrophic factor (BDNF), known to stimulate nerve growth, helps the spinal cord to recover in rats.

Wilson, Edmund B. *The Cell in Development and Inheritance.* New York: Macmillan, 1896. This book offers a late-19th-century perspective on cell biology.

Web Sites

Artificial Heart Laboratory. Available online. URL: http://www.sitemaker.umich.edu/ahl/. Accessed April 1, 2009. This Web site presents news and information on the ongoing tissue engineering efforts of the University of Michigan's Artificial Heart Laboratory.

Scripps Research Institute. Available online. URL: http://www.scripps.edu/. Accessed April 1, 2009. This Web site provides news and information on the faculty, research, and training programs at the Scripps Research Institute.

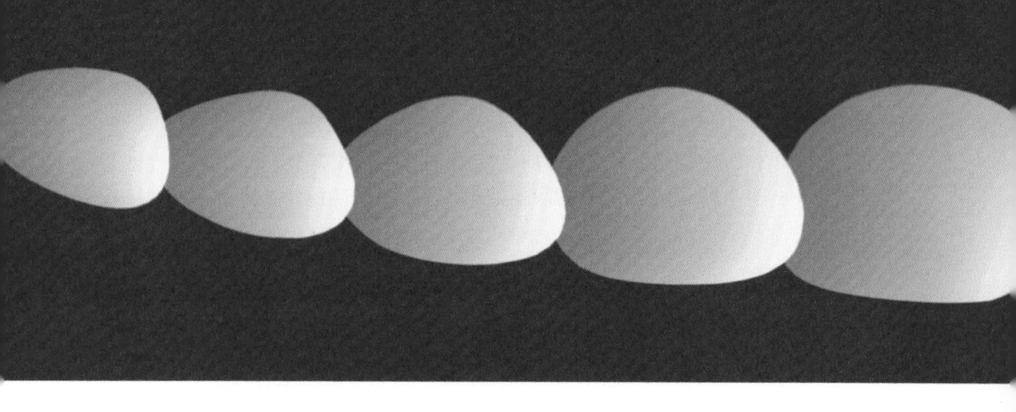

Final Thoughts

As biologists continue to make progress on the six research topics covered in this book, as well as many other important problems, a more complete understanding of organisms, behavior, and biological systems will emerge. A lot of people find this knowledge extremely satisfying. Intellectual curiosity motivates many biologists of today as much as it motivated Charles Darwin, who endured the hardships of a five-year voyage on HMS *Beagle* in the 1830s in order to study a variety of plants and animals.

But knowledge also has practical applications. Improvements in medical diagnoses and treatments are among the most visible and laudable applications of biological knowledge; brain surgeries, genetic testing, and vaccinations are just a few of the procedures that have greatly benefited, or even owe their existence to, biological research.

Other applications have barely begun, but may have no less of an impact. Several applications in particular are worth mentioning, not only because of their potential, but also because they represent some of the fears of those who worry about biology's impact—and possible encroachment—on humanity. Although these fields of study belong to some of the most remote frontiers of biology at present, they have already generated much discussion and debate, and will continue to do so well into the future.

One application is known as genetic engineering. The identification of genes and an understanding of what they do creates opportunities to intervene, particularly for those who are less than satisfied with what nature bestowed upon them. Genetic diseases are prime targets for this kind of treatment, especially those diseases that are due to a single gene defect. A technique known as gene therapy attempts to fix the defect by introducing a correct copy of the gene in the affected cells. Researchers have not had much success with this technique thus far, despite hundreds of tests; as mentioned in chapter 5, a few cases of children with a serious immune system disease were successfully treated starting in the late 1990s, but some of the patients developed leukemia as a result of the gene insertions.

While the concept of gene therapy is not contentious, the idea of tinkering with the genes of one's offspring raises much thornier issues. Germ-line engineering involves adding, deleting, or altering the genes of germ cells—the cells that generate reproductive cells such as sperm and ova. The resulting offspring inherit these alterations permanently, and because all cells of an organism arise from a single fertilized egg cell, every cell in the offspring, including the offspring's germ cells, will have them, so its children will inherit the changes as well.

Germ-line engineering opens the possibility of controlling the genes of one's children by inserting desirable genes—perhaps the same genes carried by a well known genius or a Super Bowl MVP. The children that would result from this kind of engineering are sometimes called "designer babies," referring to the design of their genetic characteristics.

Some people believe that germ-line engineering enriches human potential, while other people see it as a dangerous usurpation of the laws of nature. The debate hinges on personal values and philosophies, and is not likely to be resolved any time soon. In any case, the technology is not yet available. Biologists have been able to perform germ-line engineering in mice since the 1980s, but other species, even rats (which are quite similar to mice), are much harder to manipulate in this way for unknown reasons. Experimentation is not permissible—yet—in humans.

Other applications with a lot of promise involve prostheses—replacing or augmenting organs or tissues. The idea is an old one, going back as far as ancient Egypt (a 3,000-year-old mummy has been found with a wooden toe). Artificial limbs have greatly increased in sophistication, with the use of advanced materials such as carbon composites, and engineers have even built artificial hearts. More elaborate devices have also been considered, though not yet built, such as the right arm that gave enormous strength to the lead character in the 1970s television series *The Six Million Dollar Man* (which was based on the 1972 novel *Cyborg* by science fiction writer Martin Caidin).

If the limbs of the patient remain intact but control over them is lost, as is the case in paralysis, the result is immobility. Spinal damage, which severs or disrupts the nerves leading from the brain to the body's muscles, has confined hundreds of thousands of patients to wheelchairs or beds. Some biologists are working on mechanisms to repair or regenerate spinal nerves, as discussed in chapter 6, but other researchers are trying to design prostheses that can interpret the patient's brain signals and move accordingly. These devices will interface between the patient's brain and artificial limbs or machines able to provide substitute means of locomotion. The patient's thoughts will control the device, permitting even a completely paralyzed patient to get around.

Simple brain-machine interfaces, which connect only a few brain cells to a mechanical or electrical device, have already been tested. Cochlear prostheses, which amplify sound and directly stimulate the auditory nerves of the patient, have returned a rudimentary but nonetheless important amount of hearing to thousands of deaf patients worldwide.

Research on "reading" the thoughts of patients and translating them into movements is not quite as advanced, but Andrew B. Schwartz at the University of Pittsburgh in Pennsylvania and his colleagues have inserted slender wires into a monkey's brain with which the animal can learn to control a robotic arm. The wires pick up signals from neurons in a part of the cerebral cortex involved in movement, and after a few days of training, the monkeys can move the robotic arm using only their brain activity to retrieve a marshmallow, for which they are rewarded (they eat the marshmallow).

But while everyone applauds efforts to alleviate suffering and cure diseases, some people worry that perhaps the biological sciences are going too far. Sensory prostheses and robotic arms driven by neural networks are fine, but such advanced technology invokes unsavory images of brainwashing and mind control. An improved understanding of the brain may lead to desirable outcomes as well as not so desirable ones, such as improved ways of controlling and manipulating personal behavior. The controller or manipulator in some cases may be a ruthless dictator rather than a benign, well-meaning scientist.

Researchers have little influence on the uses, or abuses, to which their research is ultimately applied. Yet scientists are increasingly paying attention to ethics—the study of moral principles and conduct—and the future of science and technology. Five percent of each year's budget of the National Human Genome Research Institute goes to study ethical, legal, and social implications associated with research, and potential advances, in the field of human genomics.

The biological sciences have progressed a long way since the time of Pasteur and Darwin, and biologists of today are optimistic about the future. This optimism is certainly justified in the topics described in this volume, but attending these successes is a considerable amount of concern over how the knowledge will be used, and whether our descendants will be glad to have it. One of the most important frontiers of science is just beginning to be explored—the study of the consequences, intended and unintended, of scientific research.

GLOSSARY

alleles different forms or versions of a gene

amino acids molecules consisting of chemical groups called amino groups and carboxylic acid that form the units of proteins

antibodies proteins secreted by the immune system that recognize and bind to specific invaders

bacteriophages viruses that attack bacteria

bases in regards to nucleic acids, any of four different units comprising a nucleic acid sequence, which for DNA consists of adenine (A), thymine (T), guanine (G), or cytosine (C), and for RNA consists of adenine (A), uracil (U), guanine (G), or cytosine (C)

biodiversity the variety of biological organisms in an environment

biomass the weight of organisms, often used when measuring the abundance of a species in an ecosystem

blastema collection of undifferentiated cells capable of growing new tissue

capsid the protein coat surrounding and protecting a viral genome

cell the basic unit of life, consisting of biological molecules enclosed in a membrane

cerebral cortex thin but critical layers of brain tissue surrounding the cerebral hemispheres

cerebral hemispheres the right and left halves of the brain

codon a sequence of three bases in a nucleic acid that codes for a specific amino acid

conformation the shape or geometry of a molecule, especially a protein

crystal a solid in which the constituents are arranged in an orderly pattern or structure

denatured the loss of a molecule's shape or conformation

deoxyribonucleic acid a nucleic acid with deoxyribose as the sugar; these nucleic acids store the organism's genetic information

differentiate in cell biology, the process of a specialization by which a cell begins to display the specific characteristics of a certain kind of cell, such as a brain cell, heart cell, skin cell, and so forth

DNA *See* **deoxyribonucleic acid**

ecosystem an ecological system, consisting of the environment and the organisms living within it

EEG *See* **electroencephalogram**

electrodes electrical conductors, often made of metal, used to record or transmit electrical current in biological tissues

electroencephalogram a recording of electrical activity of the brain, as detected with electrodes that are usually placed at the scalp

exons the segments of a gene used for coding purposes

fMRI *See* **functional MRI**

food webs the feeding relationships, such as predation, of organisms in an ecosystem; also known as food chains

functional MRI an MRI technique used to create a series of images that detect, or map, activity of the brain

genes although there is no stringent definition of this term, the reference is generally to the units of inheritance, or the corresponding region of DNA encoding a protein that generates or affects the inherited trait

genome the genetic material of an organism

hydrophilic displaying an affinity with water molecules

hydrophobic repelled by water; such substances tend to form clumps or droplets in water

inherited traits characteristics, either physical or behavioral, that are passed from parent to offspring

intron a sequence deleted from a gene

ion channels molecules that permit the flow of charged particles called ions across the membranes of cells

isotope one of several possible forms of an atom, having the same number of protons in the nucleus but a different number of neutrons

magnetic resonance imaging a technique that takes pictures or images of the body with the aid of magnetic fields and radio waves

magnetoencephalography the process of recording the weak magnetic fields generated by the electrical activity of the brain

MEG *See* **magnetoencephalography**

metabolism the chemical reactions by which life processes such as energy production take place

motifs structural elements or shapes that tend to occur in a number of different proteins, and that may serve a similar function with the proteins in which they appear

MRI *See* **magnetic resonance imaging**

mRNA messenger RNA, a molecule that specifies the sequence of a protein

mutation a rare variation in genetic information

neoblasts undifferentiated cells that move to the sites of injury and proliferate, replacing the lost tissue

neural networks groups of neurons that make synapses with one another and process information

neuron a cell in the brain that processes information

neuroscience the study of nervous systems

NMR *See* **nuclear magnetic resonance**

nuclear magnetic resonance a set of properties displayed by atoms and molecules when placed in a strong magnetic field; scientists often use these properties to study the structure of proteins

nucleic acid an RNA or DNA molecule consisting of a chain of units called nucleotides, connected with chemical bonds

nucleotides a molecule containing a nitrogen-containing base, a phosphate group, and a five-carbon sugar

pandemic an outbreak of a disease in widespread areas

PET *See* **positron emission tomography**

pharmacogenomics the study of how genetic information affects a person's response to medications

planarians a group of flatworm species that exhibit remarkable powers of regeneration

positron emission tomography a technique to image the metabolic activity of the brain with the use of weakly radioactive materials

prion an infectious agent believed to be composed of malformed proteins

proteins an important group of molecules, involved in all aspects of an organism's structure and function, consisting of a sequence of units called amino acids

proteome the complete set of the proteins made by an organism

reassortment in viruses that have segmented genomes, the process of combining genetic segments from different viruses

regeneration the process of regrowing a lost or damaged limb or tissues

resolution the ability to discern small details in an object or an event

ribonucleic acid a nucleic acid with ribose as the sugar; these nucleic acids are involved in a variety of functions, including the production of specific proteins

RNA *See* **ribonucleic acid**

single nucleotide polymorphism a variation of a single nucleotide in a nucleic acid sequence

SNP *See* **single nucleotide polymorphism**

stem cells an important class of cells that give rise to many different kinds of cell in the body

synapse junction between neurons through which communication and information transfer occurs

transcription creation of an RNA molecule with a DNA sequence as the template

translation the production of a protein based on the sequence of the mRNA

trophic pertaining to eating or nutrition

vaccine weakened or dead infectious agent that stimulates the body's immune system, helping the body to fight off future attacks

virus an infective agent consisting of genetic material enclosed in a protein coat

X-ray electromagnetic radiation, similar to visible light except that the frequency, and energy, is much higher

FURTHER RESOURCES

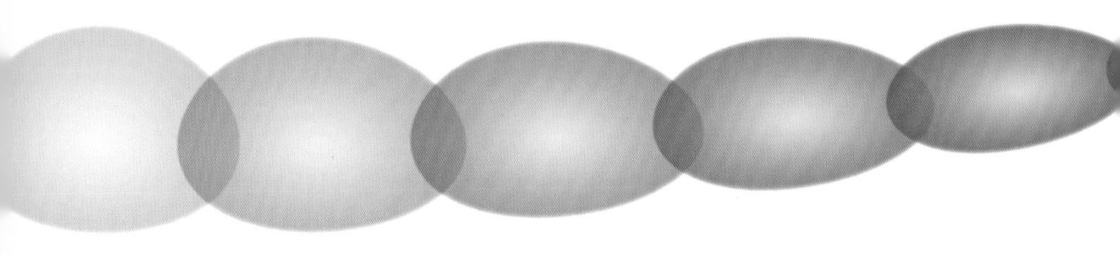

Print and Internet

Darwin, Charles. *The Origin of Species.* New York: Gramercy Books, 1995. First published in 1859, this masterpiece documents Darwin's ideas on evolution. Darwin did not get everything right, but his insight, especially considering the state of biological science in the 19th century, is astounding. The book is not light reading, but well worth the trouble, and is essential for understanding how the theory of evolution developed.

Murphy, Michael P., and Luke A. J. O'Neill, eds. *What Is Life? The Next Fifty Years: Speculations on the Future of Biology.* Cambridge: Cambridge University Press, 1997. The essays in this volume are written by a number of scientists having expertise in a variety of biological sciences as well as physics and mathematics. The authors offer their thoughts on the philosophy and future of biology, with an emphasis on human biology.

Nobel Foundation. "The Nobel Prize in Physiology or Medicine." Available online. URL: http://nobelprize.org/nobel_prizes/medicine/. Accessed April 1, 2009. Each year since 1901, the Nobel Foundation, created by Swedish inventor and innovator Alfred Nobel, has awarded prizes to recognize excellence in science and other pursuits that serve humanity. This Web page documents the winners of the Nobel Prize in physiology and medicine, including their biographies and scientific contributions.

Saul, Leif. "Biology in Motion." Available online. URL: http://www.biologyinmotion.com/index.html. Accessed April 1, 2009. Saul, a biology teacher and game designer, fills these Web pages with animations depicting basic biological processes such as evolution, digestion, and blood circulation.

Serafini, Anthony. *The Epic History of Biology.* New York: Basic Books, 2001. Starting from the beginning, this book documents the ideas of the ancient Greeks and Romans, then on through the Middle Ages to Leeuwenhoek, Linnaeus, Darwin, Pasteur, Morgan, concluding with the concepts of genetic engineering.

Thomas, Lewis. *Lives of a Cell: Notes of a Biology Watcher.* New York: Penguin Books, 1978. Thomas, the late physician and dean of Yale Medical School, first published these essays in the *New England Journal of Medicine.* Covering a broad variety of biological topics, Thomas's literary skill invites the reader to see the beauty and complexity of biology.

Watson, James D. *The Double Helix: A Personal Account of the Discovery of the Structure of DNA.* New York: Touchstone, 2001. The 1953 discovery of DNA's double helix structure ranks as one of the greatest findings in the history of biology. Watson, one of the co-discoverers (along with Francis Crick, Maurice Wilkins, and Rosalind Franklin), provides the details in this account, first published in 1968.

Weisman, Alan. *The World without Us.* New York: Thomas Dunne Books, 2007. Human civilization has left its mark on the planet, but it is not an indelible one, and it is fascinating to consider what would happen if the world was left to its own devices once again. (Which is a nice way of saying that humans become extinct.) This book describes the subsequent decay of buildings, roads, and other structures, but also discusses the effect on wildlife, which makes it an interesting exercise in biological speculation.

Wilson, Edward O. *The Diversity of Life.* New York: W. W. Norton, 1993. Wilson, an entomologist—an expert in insects—and long-time professor (now retired) at Harvard University, discusses the history of life on Earth, and why it branched out into so many different species.

Web Sites

Biology News Net. Available online. URL: http://www.biologynews.net/. Accessed April 1, 2009. Visitors to this Web site will find continually updated news on the latest biological research, as well as a discussion forum and an archive of previously featured news articles.

Exploratorium. Available online. URL: http://www.exploratorium.edu/. Accessed April 1, 2009. The Exploratorium, a museum of science, art, and human perception in San Francisco, has a fantastic Web site full of virtual exhibits, articles, and animations, including much of interest to biologists and biologists-to-be.

How Stuff Works. Available online. URL: http://www.howstuffworks.com/. Accessed April 1, 2009. This Web site hosts a huge number of articles on all aspects of technology and science, including biology.

National Institutes of Health. Available online. URL: http://www.nih.gov. Accessed April 1, 2009. The National Institutes of Health (NIH) is a United States government agency that invests billions of dollars each year in biological and biomedical research. The agency's Web site explains NIH's role and history, and contains much information on health and research topics.

Tree of Life. Available online. URL: http://tolweb.org/tree/phylogeny.html. Accessed April 1, 2009. This Web project assembles information from many contributors on a huge number of plant and animal species, including present species as well as extinct ones. The Web site contains thousands of pages, organized on the basis of phylogeny (evolutionary history)—the various branches form the "tree of life," allowing the visitor to explore the relationships among the diverse forms of life on Earth.

INDEX

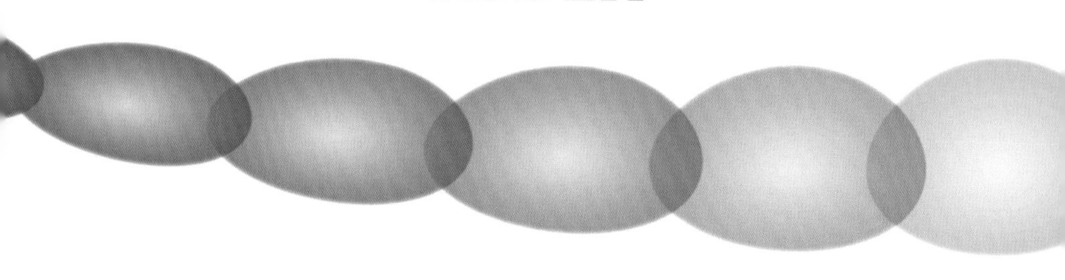

Note: Page numbers in *italic* refer to illustrations; *m* indicates a map; *t* indicates a table.

A

B

C

D

E

Q

R

S

T

U

V

W

X

Y

Z